The Bible foretold
modern man's countdown to extinction!

Many people today are trying to find a way of predicting the future. Many are even turning to astrology and witchcraft.

"I believe this generation is overlooking the most authentic voice of all, and that's the voice of the Hebrew prophets. They predicted that as man neared the end of history as we know it that there would be a precise pattern of events . . . And all of this would be around the most important sign of all—that is the Jew returning to the land of Israel after thousands of years of being dispersed."

—Hal Lindsey

(*see reverse page*)

The new national bestseller
The Late Great Planet Earth

You know, I used to come to the beach to get away from things. Just the relaxing of the waves pounding the shore. But now even the ocean is a reminder that man may be running out of time. Scientists tell us today that we are approaching a time when the ocean may not be able to sustain life anymore.

The Secretary General of the UN recently told us that man has perhaps ten years to solve the problem of survival. He pointed out the three great crises which are unique to this generation—the problem of nuclear weapons, the problem of over-population, the population explosion, the problem of pollution of our air and water. And he said that if we don't solve these problems in this decade, we are approaching the time when they will be beyond our capacity to control. Many are, for this reason, trying to find a way of predicting the future. Computer programs are being set up to try to project which might be the future of man. Many are even turning to astrology and witchcraft.

I believe this generation is overlooking the most authentic voice of all, and that's the voice of the Hebrew prophets. They predicted that as man neared the end of history as we know it that there would be a precise pattern of events which would loom up in history. Nations would fit into a certain power pattern. And all of this would be around the most important sign of all—that is the Jew returning to the land of Israel after thousands of years of being dispersed. The Jew is the most important sign to this generation.

—Hal Lindsey in the documentary film "The Return," Evangelical Communications Research Foundation producers, Pyramid Films distributors

THE LATE GREAT PLANET EARTH

by
HAL LINDSEY
with
C. C. Carlson

BANTAM BOOKS • TORONTO • NEW YORK • LONDON
A NATIONAL GENERAL COMPANY

THE LATE GREAT PLANET EARTH

A Bantam Book / published by arrangement with Zondervan Publishing House

PRINTING HISTORY

Zondervan edition published May 1970

2nd printing ...... June 1970
3rd printing .... July 1970
4th printing ..... August 1970
5th printing .. September 1970
6th printing .... October 1970
7th printing .. November 1970
8th printing ... January 1971
9th printing ... February 1971
10th printing .... March 1971
11th printing ...... April 1971
12th printing .... July 1971
13th printing .... August 1971
14th printing .. September 1971
15th printing ... October 1971
16th printing .. November 1971
17th printing ... January 1972
18th printing ... February 1972
19th printing .... March 1972
20th printing ...... May 1972
21st printing ...... June 1972
22nd printing .... August 1972
23rd printing .. September 1972
24th printing ... October 1972
25th printing .. December 1972
26th printing .... January 1973

Bantam edition published February 1973

2nd printing
3rd printing
4th printing

Unless otherwise indicated, the Scriptures quoted are taken from the Revised Standard Version of the Bible.

COPYRIGHT NOTICES AND ACKNOWLEDGMENTS

The copyright notices are listed below and on the page following which constitutes an extension of this copyright page.

Grateful appreciation is extended to the following for permission to quote from their copyrighted materials:

LIFE Magazine: From Reminiscences, *by General Douglas MacArthur,* Life *Magazine, © 1964 Time Inc.*

THE LOCKMAN FOUNDATION. *For Scriptures from* The New American Standard Bible, New Testament. *© 1960, 1962, 1963 by The Lockman Foundation.*

Published simultaneously in the United States and Canada

Bantam Books are published by Bantam Books, Inc., a National General company. Its trade-mark, consisting of the words "Bantam Books" and the portrayal of a bantam, is registered in the United States Patent Office and in other countries. Marca Registrada. Bantam Books, Inc., 666 Fifth Avenue, New York, N.Y. 10019.

PRINTED IN THE UNITED STATES OF AMERICA

INTRODUCTION

This is a book about prophecy—Bible prophecy. If you have no interest in the future, this isn't for you. If you have no curiosity about a subject that some consider controversial, you might as well stop now.

For the past thirteen years I have been giving messages on Bible prophecy throughout the United States, Canada and Mexico. The interest in this aspect of Bible study has been amazing, particularly in the past few years. People who are confronted with the state of the world today are eager to learn what the Best Seller has to say about the future. As a traveling speaker for Campus Crusade for Christ I had the opportunity to give messages on prophecy to thousands of people. These messages have consistently proven to be popular with every age group.

This is not a complex theological treatise, but a direct account of the most thrilling, optimistic view of what the future could hold for any individual. I make no claim of knowing exactly when the world is going to end. In fact, I have never taken to the hills with my possessions and loved ones to await Doomsday. I believe in a hope for the future.

We have been described as the "searching generation." We need so many answers—answers to the larger problems of the world, answers to the conditions in our nation, and most of all, answers for ourselves.

How do we know in what direction we should go? How can we separate truth from opinion? In whom can we trust?

On one side we hear that the answer to our dilemma is education. Build bigger and better schools, hire more teachers, develop a smarter generation. Has the academic community found the answers? There are many students who are dissatisfied with being told that the sole purpose of education is to develop inquiring minds. They want to find some of the answers to their questions —solid answers, a certain direction.

What do the politicians say? "We have the solutions to the problems. Elect us and we'll prove it to you."

I am not downgrading the importance of electing honest, intelligent men to positions of leadership. This is important, terribly important, but are they able to provide the answers to the basic and visceral questions of man?

Throughout history we have seen impressive strides taken by men who were stepping ahead of their time. We have seen reforms advanced from ideas generated by men of vision. And yet governments change, men falter and fall, great ideas are sometimes rejected by the shortsightedness of other men. Are we able to say that the answer is in the realm of political action?

There are other places men search for answers: philosophy, meditation, changing environment, science. Please don't misunderstand me, all of these are good if used properly. However, if we are to be absolutely honest, if we are to use our intellectual integrity, let's give God a chance to present His views.

In this book I am attempting to step aside and let the prophets speak. If my readers care to listen, they are given the freedom to accept or reject the conclusions.

Hal Lindsey

CONTENTS

We believe whatever we want to believe.
DEMOSTHENES, 348 B.C.

1

FUTURE TENSE

It was a perfect night for a party. In the warm California evening the lemon trees perfumed the patio and the flickering Tiki torches cast shadows over a lavish table. The aroma of steaks sizzling on the barbecue grill was tantalizing and we were wishing someone would give the signal to eat.

To our dismay we discovered that no one else was outside. We were alone . . . with our appetites.

Crushed in a tight circle in the overheated living room were most of the guests, waiting rather impatiently for the fortune teller to read their palms. One by one they would eagerly hold out their hands, knowing they would hear flatteries, but hoping there would be some half-truth about themselves which would give them a mental uplift—an ego-builder to tell the boys at the office or the girls at the bridge table.

And so it has been from the beginning of time. People have been obsessed with the desire to know what is going to happen in the future.

From kings and rulers to servants and slaves, rich, poor, lowly and mighty people have called upon wizards, prophets, seers, magicians, the stars and the moon for powers and signs beyond the understanding of the human mind.

Astrology originated in ancient Chaldea, in the region around the Persian Gulf. All the proud kings of Chaldea, and later Babylon, had staff astrologers to advise them concerning the future.

The Greek father of historians, Herodotus, speaks of the incredible city of Babylon and mentions the enormous ziggurat from which the astrologers tracked the stars. The ruins of several of these "observatories" called ziggurats have been uncovered in the area of ancient Babylon by twentieth-century archaeologists. Obviously astrologers were VIP's in Babylon.

The Pharaohs of Egypt imported the knowledge of astrology and magic from Babylon. Some scholars have said that the Pharaohs constructed certain parts of the pyramids to line up with the stars.

The importance or position of a man has nothing to do with his superstitious concerns. Julius Caesar was the greatest of Roman leaders, and yet depended upon "augurs" for prophetic advice. An augur is a soothsayer or fortune teller. In ancient Rome there was an official high-ranking government group, probably similar to our State Department, which was called the "Board of Augurs." The great Roman orator, Cicero, was a member of this board. However, it has been said that although Cicero took his duties seriously since he was a man of great loyalty to his country, he privately considered some of the augurs' prophecies to be absurd.

Caesar depended upon "signs in the sky" to conduct affairs of state and rule the mighty Roman Empire. As the leader of the people, he set the example for his fellow Romans to become obsessed with horoscopes.

Many strange customs have been followed in determining the course of people and nations. Decisions of marriage and journeys, wars and alliances, have sometimes been made because special court seers read certain signs of omens from the entrails of a chicken!

Starry Eyed

Many people today are discovering astrology again. Popping up on the newsstands are so many pamphlets and books about the stars that one would almost think it was a new phenomenon. The hippy musical, *Hair*, has its own staff astrologer to advise the cast on personal matters and make sure the circumstances are right for business. One of the most popular songs of the moment is "Aquarius."

Astrology is having the greatest boom in its history.

All you have to do is look in almost any newspaper to find out from one of the syndicated stargazers what your forecast is for the day. Columns on the subject now run in 1,220 of the 1,750 dailies in the United States.[1]

Writers have taken astrology and applied it to almost everything one can imagine. There is *Astrology Made Practical, Astrology Made Easy, Astrology for Everyday Living, Astrology Guide to Health and Diet,* and even *Astrology Guide to Your Sex Life.*

A few years ago Bing Crosby starred in a movie in which one of his songs contained the words, "Have you heard . . . it's in the stars . . . ?" There was something prophetic about that song!

Astrologers frequently guard their trade by predicting in generalities. However, sometimes they venture beyond vagaries and are explicit in attempting to foretell exact happenings. The result may be acute embarrassment for the astrologer.

For instance, in a national publication of 1968 an article was featured that quoted a "renowned astrologer" in her predictions about the upcoming elections in November of that year. By reading the stars this astrologer said that Rockefeller would be the next president of the United States, and either Ronald Reagan or John Lindsay would be vice-president. As for Hubert Humphrey, she stated, "I've studied his chart very carefully and do not believe he will be considered as a serious contender even for the second slot on the Democratic ticket."[2]

These political gentlemen may or may not have been disappointed by the outcome of the elections of 1968, but the astrologer was probably a bit chagrined.

On Prophecy

Two hundred years ago a distinguished English statesman, Horace Walpole, said, "The wisest prophets make sure of the event first."

Many people would probably dispute that statement and draw attention to some of our famous contemporary prophets. There was Edgar Cayce, for instance, who was called America's "sleeping clairvoyant." Cayce spoke of future events while in a self-imposed hypnotic trance. He gave personal readings to guide the vocations and health and personal relationships of hundreds of persons. However, he went beyond individuals and prophesied the reversal in Japan's attitude of friendship for the U.S. before World War II, certain battles which would take place during that war, and the increase in racial tension in America.

However, Cayce's special contribution to the realm of psychic phenomena extended beyond mere prophecy. He advised others on how to develop this psychic ability, how to interpret dreams, where to find peace, and many other subjects. His telepathic statements, made during this hypnotic sleep-state, have been indexed, catalogued, and preserved by a special research society in Virginia. Although he has been dead for more than twenty years, when we called the local library to locate some of his books we were told that they were all checked out and we would have to be put on a waiting list.

What about the popularity of Jeane Dixon? Her books and articles have become national best-sellers. She has fascinated and astounded heads of nations with her uncanny skills of perception.

President Franklin Delano Roosevelt consulted Mrs. Dixon. At the height of his concern over the world crises, and plagued by his own ill health, he sought her advice. In the company of other national leaders of the past, Roosevelt sought to find out what the future would be from someone who possessed special uncanny ability to prophesy.

Jeane Dixon told President Roosevelt that eventually the U.S. would be allied with Russia against Red China. However, she warned FDR not to give Russia anything that wasn't ours to give. The president, however, evidently did not heed this advice, for it was not long afterward that the Yalta conference was held and FDR concluded a secret agreement with Russia to give the USSR certain territory which has remained under her domination ever since.

Mrs. Dixon has made some predictions which have

had striking fulfillments. She sent word, it is reported, to the late President Kennedy prior to his trip to Dallas, warning him that he would be killed there. She also predicted several years ago that we would have a Republican president elected in 1968.

By her own admission, however, Mrs. Dixon says she is not infallible. For instance, she predicted that Red China would plunge the world into war over Quemoy and Matsu in October of 1958. (If you can't remember that far back, this did not happen.) She thought Walter Reuther would actively seek the presidency in 1964. (He didn't.)

High Spirits

Spiritualism and mysticism are weaving their spells among all age groups. We saw a handsomely illustrated magazine recently that featured "Prophecy, ESP news, psychic experiences, and spiritual healing." Provocative advertising included "Communications from the Sun World," and a new book which reveals "how they live in another dimension that is only an instant away from rapport with Earth."

It is a mystic time. Famous movie stars and wealthy socialites are traveling to the countries of the Far East to consult with "holy men." The influence of spiritualism in our popular songs, jewelry, and even clothing, is obvious. We received an invitation recently from an Indian spiritualist who says she "advises on all affairs of life." She also has a guarantee (money back?) to remove evil influences and bad luck from anyone who consults her.

The spirits are running in religious circles, also. A famous bishop, now deceased, stimulated interest in psychic phenomena by his communications with his dead son.

In churches and on college campuses mediums are receiving speaking invitations. The writer of a religious column in a Los Angeles paper recently reported, "Suffering from religious persecution and occasions of fakery in the ranks since its beginnings in the mid-1800s, the spiritualist movement in America is gaining new confidence today." [3]

In the field of higher education there are more than forty colleges which now conduct psychic research under the title of parapsychology. The interest in these subjects is growing in proportion to the increase in astrology and prophecy.

C'est la Vie

"The future" is big business. Frenchmen, for instance, spend more than a billion dollars a year on clairvoyants, gypsies, faith healers, seers and prophets. In Paris there is "one charlatan for every 120 Parisians, compared with one doctor for every 514 citizens and one priest for every 5,000." [4]

If you want to hear what the future has in store for you, perhaps you should see your travel agent for the fastest way to France. But try to avoid the need for a doctor or a priest—their services are rather scarce.

Escape

We were on the campus of a large university talking with some students in the comfortable lounge of one of the dormitories when we noticed a very attractive girl hesitate in the doorway. She stood for a few moments, her eyes glancing vacantly around the room, and then darted out as if she were being chased by demons. "What on earth do you think was wrong with her?" I asked my companions.

"Probably her vibrations were wrong," answered one young man. "She might have thought there was someone or something in here that was a threat. Who knows?"

Vibrations, spirits, stars, prophets—what an absorbing interest we have today in the unknown, the unseen and the future.

In our imaginations we long to step out of our humdrum existence and into worlds beyond. Take science fiction as an example. It fascinates us. We read books from the serious to the comic about men with miraculous powers of vision or perception. We sit with fascination before our television sets as we are transported out of the present and into the tomorrows.

Tell It Like It Will Be

In talking with thousands of persons, particularly college students, from every background and religious or irreligious upbringing, this writer found that most people want reassurance about the future. For many of them their hopes, ambitions, and plans are permeated with the subconscious fear that perhaps there will be no future at all for mankind.

This is a natural and prevalent attitude. Looking at the world today, it seems to be on a collision course. As the headline expressed it in a recent news magazine —"The World in a Mess."

People are searching for answers to basic questions. It is like the two young men who went into a book store and asked for "a book on the philosophy of truth." The clerk was flustered and very apologetic. She didn't want to discourage them, but she said they were asking for the impossible.

I wish I had been in that book store. It would have been so simple to walk over to a rather prominent display and pull out the Book that generations of readers have believed contains the "philosophy of truth." It not only contains truth, but also the great themes of peace, love, and hope, which are the desires of this and every other generation.

However, compared to the speculation of most that is called prophetic today, the Bible contains clear and unmistakable prophetic signs. We are able to see right now in this Best Seller predictions made centuries ago being fulfilled before our eyes.

The Bible makes fantastic claims; but these claims are no more startling than those of present day astrologers, prophets and seers. Furthermore, the claims of the Bible have a greater basis in historical evidence and fact. Bible prophecy can become a sure foundation upon which your faith can grow—and there is no need to shelve your intellect while finding this faith.

We believe there is a hope for the future, in spite of the way the world looks today. We believe that a person can be given a secure and yet exciting view of his destiny by making an honest investigation of the tested truths of Bible prophecy.

Men's curiosity searches past and future. . . .
T. S. ELIOT

2

WHEN IS A PROPHET A PROPHET?

What would you think of prophets today who would be willing to stake their lives on the absolute truth of their claims? They could not allow themselves errors in judgment or mistakes in the smallest detail.

There were such prophets. They were daring men, sure of the source of their faith and strong in their belief. There is a book which boldly documents the statements of these men.

They are known as the prophets of Israel and their writings have been miraculously preserved in the Bible.

Passing the Test

Are we able to trust these old prophets? What were their credentials? In a book of the Old Testament, Deuteronomy, Moses predicted that there would be many prophets who would come to the Jewish people, culminating in the final and greatest of all prophets, the Messiah.

Moses anticipated a problem. How would people know whether a prophet who claimed to speak God's message was a true prophet? A question was asked of Moses which is still being asked today. "How may we know the word which the LORD has not spoken?" (Deuteronomy 18:21).

And Moses gave the answer—the true test of a prophet: "When a prophet speaks in the name of the LORD, if the word does not come to pass or come true, that is a word which the LORD has not spoken" (Deuteronomy 18:22).

Failure to pass the test of a true prophet was a bit severe. The only grade allowed was one of 100 percent accuracy. Anything less would doom the prophet to death by stoning, which was the method of capital punishment in those days (Deuteronomy 13:1–11).

A Greek proverb says that the best guesser is the best prophet. If the prophets of Israel had been simply "guessers," they would have had to take the cruel consequences—a slow and torturous death inflicted by flying stones.

The prophets of Israel also pointed out to their people causes of current problems. This is never very popular; true prophets would not be likely to win the "Most Beloved Citizen" award in their home town.

They not only made short-range predictions which would be fulfilled during their lifetimes, but also projected long-range predictions about events far in the future. Often they didn't understand the significance of their own prophecies (I Peter 1:10–12). However, many of these prophecies are predictions of exact historical happenings which will lead to the end of history as we know it.

The astonishing thing to those of us who have studied the prophetic Scriptures is that we are watching the fulfillment of these prophecies in our time. Some of the future events that were predicted hundreds of years ago read like today's newspaper.

Let's examine some selected Bible prophecies and apply the test of truth to them.

Accurate Short-range Prophecy

Jeremiah, a prophet of Israel, was not vague in his prophecies. He told his people that Judah, the southern kingdom of Israel, would be invaded and destroyed by Nebuchadnezzar, the King of Babylon. It was not a very cheerful prophecy, and Jeremiah probably didn't receive a standing ovation.

He said that King Nebuchadnezzar would destroy the land, that it would be desolate (Jeremiah 25:9). Jeremiah was not the type to soft pedal; he prophesied that the capital city, Jerusalem, would be destroyed and that its people would lose their capacity to laugh.

After he presented this miserable picture, Jeremiah foretold that any Jewish survivors would be carried off to Babylon as slaves. To compound the complexity of these predictions, he said that the Israelites would be Babylonian slaves for precisely seventy years (Jeremiah 25:11).

At this point Jeremiah would probably have won a contest for "The Man Least Likely to Succeed" in Israel. In fact, he made Pashur, one of the religious leaders of the time, so furious that Pashur beat him and put him in stocks (Jeremiah 20:2).

Archaeology and ancient history confirm the fulfillment of Jeremiah's prophecy. Jerusalem was destroyed, Judah was laid waste, and her people were taken captive in Babylon for seventy years.

Why did the Jewish people preserve the messages of Jeremiah? These messages which foretold of their defeat and captivity were an indictment against them. It was because Jeremiah passed the test of a prophet which was established by Moses. As despised as Jeremiah was by his contemporaries, they did not dare destroy what they believed were God's words.

Jeremiah was such a bearer of sad tidings that even today we speak of complaints as "jeremiads."

An Arrow in the Armor

One of the lesser-known men of God who achieved a straight "A" in prophetic marksmanship was Micaiah. However, this was not a time when such perfection was awarded with a gold pin and guard, or a seal on Micaiah's diploma. On the contrary, he was thrown into prison and fed bread and water. What did he do to deserve such treatment?

Micaiah lived at a time when a good king, Jehoshaphat, was reigning in the southern kingdom of Judah. In the northern kingdom of Israel ruled Ahab, a wicked king.

For some reason Jehoshaphat visited Ahab, who immediately tried to persuade him to join forces against a troublesome Syrian enemy from an area called Ramoth-gilead.

Ahab gathered his own special prophets together—we might call them members of his Inner Security Council—and asked them whether he should go to battle against Ramoth-gilead or not. The 400 prophets were so soothing to Ahab that they should be called the "soothsayers" of his kingdom. They all agreed, without dissenting opinions, that Ahab should go into battle and that God would bring him through triumphantly.

But Jehoshaphat was not satisfied with the advice of these "prophets" in spite of their unanimous agreement. He said, "Is there not here another prophet of the LORD (Israel's God) of whom we may inquire?" (I Kings 22:7).

We can imagine how petulant Ahab must have been when he replied, "There is yet one man by whom we may inquire of the LORD, Micaiah the son of Imlah; but I hate him, for he never prophesies good concerning me, but evil" (I Kings 22:8).

Jehoshaphat insisted that Micaiah be consulted, so an officer was sent to bring him in. When the officer found him he tried to persuade the prophet to be a good fellow and go along with Ahab's "Inner Security Council" and tell the king what he wanted to hear. But Micaiah replied, "As the LORD lives, what the LORD says to me, that I will speak" (I Kings 22:14).

He realized as he said this that he was putting his life on the line.

When Micaiah was brought before the two kings, Ahab asked him if they should attack Ramoth-gilead. Micaiah answered in what must have been a tongue-in-cheek manner, "Sure, go on and win . . . the Lord will give it into your hands."

Ahab knew he was being ridiculed and commanded the prophet to say what was really on his mind. In answering, Micaiah showed the boldness that characterized a true prophet. First, he labeled all of the 400 "court prophets" as liars. Then he predicted that Ahab would be killed in the battle and that the Israeli army would be routed.

The chief of the false prophets—Ahab's "Chairman of the Security Council"—walked up and slugged Micaiah, probably with the same premonition of fear that made Pashur, the Israeli religious leader, take a whack at Jeremiah.

So Micaiah went to jail and Ahab went to battle. But first Micaiah challenged all the people to witness his grade on the test of a prophet. He said, "If you (meaning Ahab) return in peace, the LORD has not spoken by me" (I Kings 22:28).

Ahab wasn't going to take any chances with this pessimistic prophet, Micaiah, so he disguised himself during the big battle. But one of the enemy archers shot an arrow labeled "to whom it may concern" and in the most amazing and coincidental manner it struck Ahab in one small exposed place in his armor and killed him.

Micaiah passed the test of a true prophet. Ahab and all of his "yes men" flunked.

What Do You Say, Isaiah?

One of the most astounding prophets of all time was Isaiah, the son of Amoz. His brilliant, eloquent, even poetical predictions were made over a sixty-year span (740–680 B.C.) during the reign of four successive Judaean kings.

Isaiah proved himself a prophet by Moses' test many times. One striking incident was during the reign of King Hezekiah. In about 710 B.C. a mighty Assyrian army, numbering in the thousands and commanded by the vicious King Sennacherib, besieged Jerusalem (Isaiah 36:1, 2).

(If you stumble over some of these unpronounceable ancient names, glance at your daily newspaper and notice the tongue twisters assigned to people today.)

Sennacherib was a crafty fellow. He sent his eloquent "mouthpiece," a fellow named Rabshakeh, to the people of Jerusalem with a powerful propaganda speech. Rabshakeh, an ancient "Minister of Misinformation," told the people of all the defeats suffered by neighboring countries. This was brainwashing in the craftiest sense. The techniques used were intended to induce the besieged citizens of Jerusalem to surrender without a fight.

When King Hezekiah heard how his people were being intimidated, he sent a delegation to Isaiah to plead with him to pray to Jehovah, the God of Israel. Isaiah then made some short-range predictions. He said that a rumor would reach King Sennacherib of internal trouble in his kingdom and that he would return without attacking Jerusalem. Furthermore, he said that Sennacherib would be assassinated in his own land.

It happened just as Isaiah predicted. Jerusalem was spared a certain defeat and King Sennacherib was killed by his own sons (Isaiah 37:36–38).

Isaiah Predicts One Hundred Years in the Future

Isaiah also predicted that Babylon would completely destroy Judah and carry away all the treasures of Israel. He foretold that the surviving sons of the royal family would be eunuchs in the palace of Babylon. This occurred just a little over one hundred years later (Isaiah 39:5–7).

One Hundred and Fifty Years in the Future

Isaiah gave an incredible prediction when he foretold that the mighty invincible Babylonians would be conquered and so completely destroyed by the Medes that Babylon would never be inhabited again (Isaiah 13:17–22).

This was an astonishing prediction at that time since the city of Babylon became one of the seven wonders of the ancient world and was considered impregnable. Yet approximately 150 years after Isaiah's prediction the Medes and Persians besieged the towering walls of Babylon. This was no small feat. Those walls were 150 feet high and so thick that five chariots could run abreast around the top.

The Medes and Persians were cunning. They dammed the Euphrates River which ran under the Babylonian wall and through the city. While Babylon was celebrating at a drunken royal ball, the Median army marched along the dry river bed under the wall and conquered the city. This was the very night that "the handwriting on the wall" appeared to the arrogant Babylonian king. It read, "Mene, mene, tekel, upharsin." Daniel, the prophet, deciphered it to the drunken mob as follows: "Your kingdom has been weighed in the balance and found wanting. It is divided and given to the Medes and Persians." That night the Babylonian kingdom fell (Daniel 5:1–31).

Two Hundred Years in the Future

Isaiah predicted that a certain king named Cyrus would see that Jerusalem and the Temple were rebuilt by allowing those who wished to do this work to return to the land of their ancestors (Isaiah 44:28—45:4).

Almost 200 years after this prediction a certain Persian king, also named Cyrus, granted royal favor to the Jewish captives left over from Babylon and sent them to Jerusalem with a requisition for materials to rebuild the city (Ezra 1:1–11).

The Prophet Preserved

Many so-called Biblical scholars today try to "late date" such predictions as Isaiah's to make his prophecies seem to be after the fact. To do this not only violates the consistent witness of the history of those times, but also makes the Jewish people religious charlatans and deceivers. The Jews would have had no reason to keep for posterity those writings of the prophets if they were a fraud.

The Israeli prophets were not popular, as we have shown. In the times of Jeremiah, Micaiah, Isaiah, and the others, the odds were against anything they wrote or said being accepted. Yet they were preserved above all the other writings of their day.

What made these prophets so special? The answer lies in the test of a true prophet which Moses gave—their prophecies must all come true. They could not be ignored. We cannot ignore them. They passed the test —summa cum laude.

History teaches us that man learns nothing from history.

HEGEL

3

DO WE REALLY LIVE AND LEARN?

It's ironic that man never seems to learn from past mistakes, especially when they relate to major catastrophes. World War I was called the war to end all wars, yet within a generation World War II was fought in basically the same arena. We are now running around the world desperately seeking to put out fuses which could explode into what might be the last war on earth.

Through the grim pages of history we see the record of man's constant struggle to live with his fellow man. Families fight against families, tribes against tribes, and nations against nations. Most people hate war, and yet since recorded time the world has seldom seen peace.

General Douglas MacArthur said, "Men since the beginning of time have sought peace . . . military alliances, balances of power, leagues of nations, all in turn failed, leaving the only path to be by way of the crucible of war. The utter destructiveness of war now blots out this alternative." [1]

Mankind has not learned the futility of war from history. However, as tragic as this is, there is another lesson, even more tragic, which has not been heeded. This involves a people whose most cherished hope had been the coming of their great Deliverer, called the Messiah.

The central theme of the Jewish prophets was that "the Messiah" would come and fulfill the promises given to their forefathers, Abraham, Isaac, and Jacob. In these promises Israel is to be the leading nation of the world under the reign of the Messiah who would bring universal peace, prosperity, and harmony among all peoples of the earth.

The paradox is that there has come a Jew who claimed to be the Messiah. He fulfilled many of the ancient predictions, but was rejected by those who should have recognized Him first. The question is this: If He was truly the long-awaited Messiah, as millions have believed, why didn't the majority of the religious leaders of His day believe His claims? These religious leaders, after all, knew the Messianic predictions. Ignorance was not their excuse. The reasons they didn't believe Him are fascinating and extremely relevant to this hour of history in which we live.

Two Portraits

Two completely different portraits of a coming Messiah were described by the Old Testament prophets. The portraits, painted by the sure hand of God, were placed on the same canvas, framed in one picture.

For those who lived prior to the birth of Jesus of Nazareth, the perspective of these two portraits of the Messiah was difficult to understand.

Imagine a man looking at a range of mountains. He is able to see the peak of one mountain, and beyond it the peak of another. However, from this vantage point, he cannot see the valley which separates these two mountains.

Men viewed the two portraits of the Messiah in the same manner. They saw two different persons, but missed the connection. They did not perceive that there could be just one Messiah, coming in two different roles, and separated by the valley of time.

One portrait of the Messiah depicts Him as a humble servant who would suffer for others and be rejected by His own countrymen. This portrait we may call "the Suffering Messiah." (Look into the prophecies of Isaiah 53 for the perfect picture of this Messiah.)

The other portrait shows the Messiah as a conquering king with unlimited power, who comes suddenly to earth at the height of a global war and saves men from self-destruction. He places the Israelites who believe in Him as the spiritual and secular leaders of the world and brings in an age free of prejudice and injustice. It's easy to see why this would be the most popular portrait.

We may call this second picture "the Reigning Messiah." We find this description in such prophecies as Zechariah 14 and Isaiah 9:6, 7.

These two portraits of the Messiah presented such a paradox that the rabbis at least a century before Jesus of Nazareth was born theorized that there would be two messiahs. They could not see how both portraits could be true of the same person. Their misunderstanding led them to believe that the suffering one would deliver the people from their sins by bearing the penalty of death for them. He would be primarily a "Spiritual Deliverer." The reigning king would conquer Israel's enemies and bring world peace. He would be primarily a "Political Deliverer."

The Big Question

Why did the majority of the Jewish people, who knew the teachings of their prophets, reject Jesus of Nazareth as their Messiah when He came? Why did they ignore the portraits of the suffering Messiah? Jesus Himself pointed out to them Old Testament predictions concerning His life and ministry which were being fulfilled in His life.

First of all, the Jews didn't take their prophets literally as far as this suffering Messiah was concerned. They took very literally the portrait of the Messiah who would come as the reigning King. However, they had degenerated in their own religious convictions to the point where they didn't believe they were sinful. They believed they were keeping the law of Moses; therefore, they saw no need for a suffering Messiah to deliver them from their sins.

The Jewish religious leaders had built up a tradition of interpretation which made keeping God's laws merely an external thing (Mark 7:1–15). Jesus, however, pointed out the true meaning of God's law in the Sermon on the Mount. He showed that murder was not just actually killing someone, but that it was being angry with your neighbor without just cause (Matthew 5:21, 22). He also pointed out that adultery in God's sight was merely looking at a woman with lust (Matthew 5:27–32).

He expounded the real meaning of "love your neighbor as yourself" as applying to your enemies as well (Matthew 5:43–48).

With each commandment Jesus emphasized the false interpretation of the religious leaders and contrasted it with the true meaning which God had always intended. He showed that God looks upon the heart and not just upon the outward performance of man.

I am sure that if you took the previous paragraph seriously you may feel a little uncomfortable right now. You may be saying to yourself what I said some years ago when the true meaning of the Ten Commandments was first pointed out to me. "Who in the world can be accepted by God if he is going to have to keep the law in his thoughts and motives?"

If this is what you are thinking, congratulations! You have just discovered the whole purpose for the law of Moses. The commandments were never intended to be used to work our way to God. The commandments were primarily given to show us how perfect we would have to be in order to earn acceptance with God by our own good deeds. This is why God says to us, "Whoever keeps the whole law and yet stumbles in one point, he has become guilty of all" (James 2:10 NASB).

It is easy to see in the light of this why the Bible says, ". . . by the works of the Law no flesh (man) will be justified in His (God's) sight" (Romans 3:20 NASB).

The law was given to show mankind why it needed a "suffering Messiah" who alone could make man acceptable to God. Any person who hasn't come to see that his most basic problem is an inner spiritual one prefers a political deliverer to a spiritual one. It is not difficult, therefore, to understand the basic attitude which rationalized away the prophetic portrait of the suffering Messiah.

Jesus presented His credentials as the suffering Messiah, but many rejected Him because they were looking for a great conqueror. They were looking for that political leader who would deliver them from the Roman oppression. In their blindness they discounted more than *300 specific predictions* in their own sacred writings about this Messiah.

The second reason why the Jewish people rejected the Messiah was one of indifference—an indifference to their spiritual need. They couldn't be bothered. They were too busy. They had a chariot race or a party to attend. Caught in the treadmill of daily existence, the deeper needs of the inner self weren't important.

In addition, they didn't bother to do any investigation for themselves. Many knew there was something unusual about this carpenter from Nazareth, but their religious leaders rejected Him and they took their opinions instead of searching for the truth themselves.

It was because people will not do their own research that Jesus made an astounding statement on the signs of the times. He showed simply and clearly how prophecies were being fulfilled by His life. And yet, to the sorrow of many, He was ignored.

Signs of the Times

The religious leaders of His day were the number one skeptics. They came to Jesus and asked Him to show them a sign from heaven. They wanted some sensational miracle which would give them proof that Jesus was their promised Messiah. They wanted Him to suddenly step out of the sky as the conquering Messiah (as He is revealed in Zechariah 14) and take over all the kingdoms of the world and defeat the Roman Empire.

Jesus had already given them important signs to prove who He was. He had healed many persons and had raised at least one from the dead. But they didn't consider this sufficient evidence to prove His claim of being the Messiah.

Jesus answered these religious leaders by saying: "When it is evening, you say, 'It will be fair weather, for the sky is red.' And in the morning, 'It will be stormy today, for the sky is red and threatening.' You know how to interpret the appearance of the sky, but you cannot interpret the signs of the times" (Matthew 16:2, 3).

The signs of the times. It is important to see what Jesus was driving at. People in Palestine, even today, indulge in being amateur weather forecasters. Weather conditions are such that the particular signs given here are fair indications of what the weather will be.

Jesus said that the signs leading up to His coming were just as clear as the face of the sky. Let's examine these signs, these credentials. Let's look at specific predictions of how this man Jesus would come to fulfill the role of the Messiah, of how He would fit into the first portrait of the suffering Messiah.

The first theme of predictions relates to the circumstances of His birth. These were His credentials.

Birth Credentials: His Family

God revealed to Abraham, the father of all Jews, that he would have a direct descendant who would be a blessing to all the peoples of the earth (Genesis 12:1–3).

Furthermore, God revealed to Jacob, who was one of the descendants of Abraham, that the Messiah would come through one specific tribal state—the tribal state of Judah. The Jews established states in the land of Palestine after they took it over, and they were divided according to twelve families. These families originated from the twelve sons of Jacob (Genesis 49:10).

His family credentials were narrowed even more, from the tribe of Judah to the family of David. The prediction was given to David, the great king in the history of Israel, by the prophet Nathan.

Nathan said, "Thus says the Lord of hosts . . . I will raise up your offspring after you, one of your own sons, and I will establish his kingdom. He shall build Me a house, and I will establish his throne for ever. I will be his father, and he shall be My Son . . ." (I Chronicles 17:11–13 Amplified).

King David was promised at least two exciting things here. First, one of his direct descendants would reign forever. Secondly, this person would not only be one of his direct descendants, but in a mysterious way He would be uniquely the Son of God.

Rabbinic tradition ascribed this to be a prediction of

the Messiah. Consequently, one of the most common Messianic titles is "the Son of David."

Birth Credentials: The Place

The prophet Micah lived seven hundred years before the birth of Christ. He was a contemporary of the great prophet Isaiah. It was revealed to Micah that the Messiah would be born in Bethlehem.

"But thou, Bethlehem Ephratah, though thou be little among the thousands of Judah, yet out of thee shall he come forth unto me that is to be ruler in Israel; whose goings forth have been from of old, from everlasting" (Micah 5:2 KJV).

This was a prophecy of the Messiah which was unmistakable because it referred to His eternal pre-existence . . . "whose goings forth have been from of old, from everlasting."

This was no ordinary man, but a supernatural person who would invade history from Bethlehem.

This prophecy is quoted in the New Testament in Matthew, in answer to Herod's question to the Hebrew theologians about where the Messiah would be born. They answered him: "In Bethlehem of Judea, for so it is written by the prophet" (Matthew 2:5).

How remarkable it is that century after century the Hebrew people handed down these detailed prophecies. They had to be revelations from God, otherwise they would not have been preserved in such a consistent manner.

Birth Credentials: The General Time

We have examined the prophecies concerning the family line of the Messiah and the place of His birth. Let's look at the time factor.

The prophet Daniel, while in captivity in Babylon, was given a precise timetable and sequence relating to the future events of the people of Israel. Daniel was told that there would be a certain number of years which would transpire between the time a proclamation was given which allowed the Jewish people to return from their Babylonian captivity back to Israel and the coming of the Messiah.

This proclamation can be established according to Scriptural history in Nehemiah 2:1–10. Also, archaeologists have uncovered evidence of this same proclamation in the ancient Persian archives.

From the time permission was given to return and rebuild the city of Jerusalem and the Temple until the Messiah would come as the Prince, the heir-apparent to the throne of David, would be 483 years (69 weeks of years—483 years). Sir Robert Anderson of Scotland Yard spent many years of his life verifying and validating the details of this prophecy. He wrote a comprehensive book of his study called *The Coming Prince.*

Not only was Daniel given specific years, but also a sequence of major historical events which cannot be denied.

First of all, there was the proclamation given to the Jews to return from captivity and rebuild the Temple.

After that, the Messiah would come as the Prince.

Then the Messiah would be "cut off," which is an idiom for being killed.

After the Messiah was killed an army would sweep in and destroy the city and the Temple which was rebuilt previously by the returned Babylonian exiles (Daniel 9).

Daniel's prophecy shows that whoever the Messiah was He had to appear before the city and the Temple were destroyed in A.D. 70 by Titus of Rome.

There was only one person who was taken seriously as the Messiah before A.D. 70. We have the logical candidate for that role in the carpenter from Nazareth.

Credentials Relating to His Ministry

We have considered a panorama of Jesus' birth credentials to verify His claim as the Jewish Messiah. To pile proof upon proof, consider the prophecies relating to the deeds Jesus would do upon earth—prophecies about His own ministry.

The prophet Isaiah presented a vivid description of the coming Messiah when he said: "Say to those who are of a fearful heart, 'Be strong, fear not! Behold, your

God will come with vengeance, with the recompense of God. He will come and save you.' Then the eyes of the blind shall be opened, and the ears of the deaf unstopped; then shall the lame man leap like a hart, and the tongue of the dumb sing for joy. For waters shall break forth in the wilderness, and streams in the desert" (Isaiah 35:4–6).

Jesus knew the Old Testament prophets. He quoted this exact prophecy from Isaiah when John the Baptist seemed to have some doubts about Him. John was the herald who announced the coming of Jesus as the Messiah, and yet even John could not reconcile the two portraits of the Messiah. When he was taken prisoner he sent some messengers to Jesus to ask, "Are You the Coming One, or shall we look for someone else?" (Matthew 11:3 NASB).

Jesus answered by quoting the predictions of the miracles He would perform. "Go and report to John the things which you hear and see: the blind receive sight and the lame walk, the lepers are cleansed and the deaf hear, and the dead are raised up, and the poor have the gospel preached to them" (Matthew 11:4, 5 NASB).

The fact that Jesus was performing these very miracles was His credentials to substantiate His claim that He was the Messiah.

Credentials Relating to His Rejection and Suffering

Probably the most phenomenal predictions relating to the suffering Messiah's portrait are the prophecies that show His rejection and suffering. One of the great passages on His rejection is Isaiah 53. This is called the "bad conscience of the synagogues" because it is no longer read in the temples on holy days, as it once was.

In Isaiah 52 there is a general look at this One who is called the servant of God. It is obvious here that the prophet is not talking about Israel, which in some passages is called the servant of God, but about one who would save Israel. "Behold, my servant, (Christ) shall prosper, he shall be exalted and lifted up, and shall be very high. As many were astonished at him—his appearance was so marred, beyond human semblance, and his form beyond that of the sons of men" (Isaiah 52: 13, 14).

This was a reference to what happened at the trials of Jesus when He was repeatedly hit in the face. "So shall he startle many nations; kings shall shut their mouths because of him; for that which has not been told them they shall see, and that which they have not heard they shall understand" (Isaiah 52:15).

In this passage the prophet is saying that He would startle, or astound nations (meaning the non-Jews) and that they would see things they had never seen. The Gentiles, in other words, would begin to understand the ways of God.

However, Isaiah speaks of His rejection by the Jews, which in itself was remarkable since Isaiah was a Jewish prophet who wrote at least 700 years before Christ was born. Isaiah predicted that his people would reject the very one for whom they looked (Isaiah 53:1–3).

Notice an interesting fact. The prophet writes of this event in the past tense, which was a common literary device of the Jewish writers . . . ("many were astonished . . . his appearance was so marred"). When they wished to emphasize the certainty of a prophecy they would put it in this tense, which is called the prophetic perfect tense in Hebrew.

It was further prophesied here that the Jews would reject this Man because He didn't have the royal splendor they desired. Isaiah said He would be "despised and rejected," which is precisely what happened.

In this part of the prediction the credentials of the Messiah as a person who would be a substitute for the iniquity, or wickedness, of man are presented: "Surely he has borne our griefs and carried our sorrows; yet we esteemed him stricken, smitten by God, and afflicted. But he was wounded for our transgressions, he was bruised for our iniquities; upon him was the chastisement that made us whole, and with his stripes we are healed. All we like sheep have gone astray; we have turned every one to his own way; and the LORD has laid on him the iniquity of us all" (Isaiah 53:4–6).

Rabbis since the birth and death of Jesus of Nazareth

have reinterpreted this passage to say that the third person singular pronoun does not refer to a personal Messiah, but to the nation of Israel. However, the passage speaks of this person as bearing the consequences of the transgressions of Israel. Israel could not be a substitute for itself, since the passage clearly states that the Lord hath laid on *Him* the iniquity of *us*.

Isaiah 53 continues to say that this person would not receive the true justice of the Jewish law. This, of course, was true during the trials of Jesus. They were astounded that He didn't try to defend Himself . . . "He was oppressed . . . yet he opened not his mouth" (Isaiah 53:7).

Again there is a clear prediction that the Messiah would die for the transgressions of Isaiah's people, as well, of course, as for the whole world. ". . . he was cut off (killed) out of the land of the living, stricken for the transgression of my people" (Isaiah 53:8b).

The details of prophecy relating to this portrait of the Messiah are exact. Isaiah says that He will die beside criminals. "And they made his grave with the wicked." His grave would be with the rich: "And with a rich man in his death" (Isaiah 53:9).

Did it happen? Of course. Jesus was crucified between two thieves. After His death one of the rich Pharisees, Joseph of Arimathaea, took pity on Him and buried Him in his own tomb. Joseph was the rich man predicted here.

We can picture Isaiah, standing at Calvary, looking at the panoramic view that Jesus saw and experienced on the cross. However, Isaiah saw this 700 years before Jesus was born!

To continue in this remarkable passage, Isaiah speaks of the fact that men could be declared righteous and acceptable to God because He bore their sins. In Biblical vernacular, this is the meaning of "justified."

Isaiah said that He "made intercession for the transgressors" (Isaiah 53:12). Most persons with only a slight exposure to Christian teaching will remember the words of Jesus from the cross when He said, "Father, forgive them; for they know not what they do."

Thirty Pieces of Silver

Another Old Testament prophet was Zechariah, who wrote almost 500 years before Jesus lived. In his book he gave another specific and minute prediction which could only refer to one person. He wrote: "And I said unto them, If ye think good, give me my price; and if not, forbear. So they weighed for my price thirty pieces of silver. And the LORD said unto me, Cast it unto the potter: a goodly price that I was prised at of them. And I took the thirty pieces of silver, and cast them to the potter in the house of the LORD" (Zechariah 11:12, 13 KJV).

Notice the three specific trends in this passage. First, there would be a time when the people would estimate their own God's worth at thirty pieces of silver. Then, these thirty pieces would be cast down in the house of the Lord, which is the Temple. Finally, the money would be given to the potter for the graves of the poor people.

Do you suppose any present-day "prophets" would dare make such an exact prediction?

It happened exactly as told. Matthew records this when he tells of Judas going to the chief priests who were plotting to murder Jesus and saying, "What will you give me if I deliver Jesus to you?" And the priests decided they would give thirty pieces of silver for the betrayal (Matthew 26:14, 15).

After Jesus was betrayed and Judas saw that Jesus was condemned to death, which was more severe than Judas had anticipated, he regretted what he had done. He went to the priests and tried to return the money, but the priests insulted him. Judas became infuriated. He threw the money down in the Temple (Matthew 27:3–5). (Prediction fulfilled.)

The priests took the money and piously said it wasn't proper to return to the treasury the price of the betrayal, so they decided to give it to the potter to buy a potter's field (Matthew 27:6–10). (Trying to ease their consciences, no doubt.)

Notice one very important point. Jesus had no control over this prophecy. It had to be fulfilled without any interference on His part. This fact explodes the major premise of a book which has gained some popularity, called *The Passover Plot*. While the writer of this book does a service by accepting the historical reality of Jesus, he claims that Jesus deliberately plotted to fulfill the predictions of the Messiah given in the Old Testament. This theory could not be valid because there is no way to explain how many predictions, such as the one about the thirty pieces of silver, could be fulfilled when the circumstances were out of Jesus' hands.

Predictions Relating to the Crucifixion

Jesus told His disciples that there were predictions of His suffering in the Psalms (Luke 24:44–46).

One of the clearest prophecies is found in Psalm 22, which was written by King David more than 1000 years before Christ. David describes events which could not have happened to himself since they were beyond the scope of his own experience.

The Psalms were accepted as the word of God and David was speaking "in the Spirit," as the ancient rabbinical schools recognized. The psalmist gives a detailed and precise prediction of a person being crucified. He speaks of the suffering of the Messiah as if he were on the cross with Him, feeling His pain, seeing the people and events around him. Speaking in the spirit of the Messiah, David says, "I am poured out like water"—he is speaking of the profuse perspiration of one hanging in the intense sun. "And all my bones are out of joint" —this is one of the most excruciating aspects of crucifixion. The ligaments stretch and the bones pop out of joint.

He tells of the intense thirst—"and my tongue cleaveth to my jaws." Jesus said on the cross, "I thirst."

"For dogs have compassed me: the assembly of the wicked have enclosed me: they pierced my hands and my feet. I may tell all my bones: they look and stare upon me. They part my garments among them, and cast lots upon my vesture" (Psalm 22:16–18 KJV).

"Dog" was a common slang expression the Jews used for the Gentiles—Jesus was surrounded by Gentiles at the crucifixion. He was crucified in the nude; this passage speaks of the shame of it. At the foot of the cross the soldiers gambled (or cast lots) for His robe.

As perfect as this passage is in its prophetic accuracy, it gains additional importance when we realize that crucifixion as a way of punishment was not known at the time David wrote this. The Jews of that time executed by stoning. It was not until about 200 B.C. when the Romans adopted this cruel practice that crucifixion was widely used—800 years after this prophecy.

Guaranteed Accuracy

If there is one thing that guarantees the historical accuracy of what the New Testament authors wrote it is the animosity of the Jewish people who crucified Jesus. The message of these prophecy fulfillments was spread by word of mouth all over the Palestinian area starting fifty days after these events happened.

If those who crucified Jesus could have disproved any of the historical realities of these events they would have done so and destroyed the whole movement from the beginning. But they didn't bring up any refutation of the facts of fulfilled prophecy; instead they put to death the persons who were proclaiming these facts.

That generation did not take seriously the credentials of the suffering Messiah. Jesus predicted the destruction of those who put Him on the cross. "For the days shall come upon you when your enemies will throw up a bank before you, and surround you, and hem you in on every side, and will level you to the ground and your children within you, and they will not leave in you one stone upon another, because you did not recognize the time of your visitation" (Luke 19:43, 44 NASB).

Was this prophecy fulfilled? As previously mentioned, Titus and the Roman legions swept down upon Jerusalem and destroyed it in A.D. 70.

Will We Learn?

Will we repeat history? Will we fail to take the prophets literally and seriously? Will we be indifferent? Will we allow those who claim to be religious leaders to explain these things away and not investigate for ourselves?

There are many more predictions about the reigning Messiah who is *yet to come* than there were about the suffering Messiah. Will we fail to weigh these prophecies for ourselves, in spite of what others may say?

The remainder of this book will present the prophecies which are related to the specific pattern of world events which are precisely predicted as coming together shortly before the coming of the Messiah the second time—coming in power to rule the earth.

Many of these predictions were in the same paragraphs as those relating to the first coming of the Messiah. Do we dare allegorize away the meaning of these?

Will these predictions be fulfilled just as certainly and graphically as those of the first coming?

This writer says positively, "Yes."

There are few countries which have played so central a role in world history as the Land of Israel.

DAVID BEN-GURION, 1965

4

ISRAEL, O ISRAEL

Some time in the future there will be a seven-year period climaxed by the visible return of Jesus Christ.

Most prophecies which have not yet been fulfilled concern events which will develop shortly before the beginning of and during this seven-year countdown.

The general time of this seven-year period couldn't begin until the Jewish people re-established their nation in their ancient homeland of Palestine.

Keys to the Prophetic Puzzle

A definite international realignment of nations into four spheres of political power had to occur in the same era as this rebirth of Israel. Each sphere of power had to be led by a certain predicted nation and allied with certain other nations. The relationships of all these factors to each other is easily determined by the following clues: first, each one of the four spheres of political power is said to be present and vitally involved with the reborn state of Israel.

Secondly, each one of these spheres of power is a major factor in the final great war called "Armageddon," which is to be triggered by an invasion of the new state of Israel.

Third, each one of these spheres of power will be judged and destroyed for invading the new state of Israel, by the personal return of the Jewish Messiah, Jesus Christ.

It should be obvious that these predicted movements

of history are interrelated in their general time of beginning and ending. This is why the prophecies can be pieced together to make a coherent picture, even though the pieces are scattered in small bits throughout the Old and New Testaments.

Raised Eyebrows

Many Bible students in recent years tried to fit the events of World War I and II to the prophetic signs which would herald the imminent return of Christ. Their failure discredited prophecy.

The people who have fled to the mountains to await the end of the world haven't had the faintest idea about the truths in Bible prophecy.

It is because of these unscriptural attempts at calculating dates that some eyebrows rise when we speak of Bible prophecy today.

Dream and Reality

The one event which many Bible students in the past overlooked was this paramount prophetic sign: Israel had to be a nation again in the land of its forefathers.

Israel a nation—a dream for so many years, made a reality on 14 May 1948 when David Ben-Gurion read the Declaration of Independence announcing the establishment of a Jewish nation to be known as the State of Israel.

In 1949, Prime Minister Ben-Gurion said that Israel's policy "consists of bringing all Jews to Israel . . . we are still at the beginning."[1]

What Countdown?

This seven-year period we have called the "countdown" is a period of unique events. There is more prophecy concerning this period than any other era the Bible describes.

The apostle John counted out seven years for this period when he spoke of the second half being forty-two months (i.e., 3½ years), and the first half being 1260 days (i.e., 3½ x 360 days, which is the Biblical year) (Revelation 11:2, 3).

The prophet Jeremiah spoke of the time when God would return His people of Israel and Judah from a great captivity and dispersion. He calls this period "the time of Jacob's trouble."

As Christ told of the world conditions that would immediately precede His coming, He said, "For then there will be great tribulation [affliction, distress and oppression] such as has not been from the beginning of the world until now, no, and never will be. And if those days had not been shortened, no human being would be saved" (Matthew 24:21, 22).

In other words, this period will be marked by the greatest devastation that man has ever brought upon himself. Mankind will be on the brink of self-annihilation when Christ suddenly returns to put an end to the war of wars called Armageddon.

Israel, the Fuse of Armageddon

What has happened and what is happening right now to Israel is significant in the entire prophetic picture. Men who have studied events that were to occur shortly before the great holocaust known as Armageddon are amazed as they see them happening before their eyes.

Too few Biblical scholars pay any serious attention to the proven prophetic content of Scripture. Dr. William F. Albright, eminent archaeologist and professor of Semitic Languages, noted this fact after he had verified many historic fulfillments of Bible prophecy. He said, "That the prophets were not only dedicated men, but also predictors of the future, is fully recognized in Biblical tradition but has been under-emphasized by modern Biblical scholars. . . ."[2]

The nation of Israel cannot be ignored; we see the Jews as a miracle of history. Even the casual observer is amazed how the descendants of Abraham, Isaac, and Jacob have survived as a distinct race in spite of the most formidable odds. What other people can trace their continuous unity back nearly 4,000 years?

Twice the Jews have been destroyed as a nation and

dragged away as slaves under inhuman circumstances; twice the Jews returned to their ancient homeland and re-established their nation.

What other people have preserved a distinct and separate national identity in spite of a total of some 2600 years of being scattered—years of dispersion, as it is called? During all these years these men and women without a country have suffered the most insane and unjust persecutions ever endured by any collection of people or nationalities.

Jewish survival is a phenomenon. However, Jewish history, with all of its tragedies and triumphs, has been accurately foretold.

Setting the Stage

History may be dry bones to some, but the history of how the stage was set for the rebirth of Israel is fascinating. This history also serves as a standard to test how reliable future prophecy concerning Israel will be. Let's look at the record of past prophecy fulfilled concerning Israel.

God's Woodshed

Some 3500 years ago, at a time when the Jewish people were enroute from Egypt to possess the Promised Land of Palestine, Moses predicted that they would be chastened, or disciplined, twice, as a nation, for not believing their God and rejecting His ways. The first stage of this disciplinary action came from Babylon.

Misery in Babylon

Moses predicted that a mighty nation would invade and destroy Israel. The invaders would be so fierce that they would have no respect for either the very old or the very young. Civilians would be massacred and property completely destroyed. The survivors would be taken as slaves.

As previously noted, the prophet Isaiah added details to what Moses predicted, about 150 years before it occurred. Isaiah said to a king of Judah, named Hezekiah: "Behold, the days are coming when all that is in your house, and that which your predecessors have stored up till this day, shall be carried to Babylon. Nothing shall be left, says the Lord" (Isaiah 39:6 Amplified).

The prophet Jeremiah, several years before it happened, predicted how long the Babylonian captivity would last: "And this whole land shall be a waste and an astonishment, and these nations shall serve the king of Babylon seventy years" (Jeremiah 25:11 Amplified).

Precisely as predicted, the Babylonians swept into the southern kingdom of Israel and Jerusalem and destroyed it. Those who survived the holocaust were carried off to Babylon as slaves where they remained for seventy years (II Chronicles 36:15–21).

At the end of this period of enslavement the Persian king, Cyrus, released some of the Jewish people to return and rebuild the Temple in Jerusalem (II Chronicles 36:22, 23). You will recall that this king had been predicted by name some 200 years before by Isaiah (Isaiah 44:28; 45:4).

The Winds of Rome

In the same prophecy in which he predicted the first stage of divine discipline, Moses also predicted the second stage. He said that because of continued disbelief and rejection of their God, Israel would be destroyed as a nation a second time. This time the survivors would be scattered throughout the world in every nation. They would be relentlessly persecuted—men without a country. Moses gives a graphic portrayal of the history of the Jews when he says that they will be scattered among all the people of the earth, where they "shall find no ease." The Jews will find no rest for their feet, their very lives shall "hang in doubt" (Deuteronomy 28:64–68).

Many other prophets, such as Isaiah, Jeremiah, Ezekiel, and Amos to name a few, all predicted the great world-wide exile of the Jewish people and the destruction of the Jewish nation.

Just before His arrest and crucifixion, Jesus said,

". . . for there will be a great distress upon the land, and wrath to this people, and they will fall by the edge of the sword, and will be led captive into *all the nations* . . ." (Luke 21:23, 24 NASB).

It's very important to note that this great theme of prophecy concerning the global dispersion was predicted by Jesus as occurring to the same generation that crucified Him. He said, "Truly I say to you, all these things shall come upon this generation" (Matthew 23:36 NASB).

History verifies the accuracy of these prophecies. Just as predicted by Jesus, less than forty years after His death Titus and the Roman legions destroyed Jerusalem and the nation, slaughtering hundreds of thousands. Those who survived were shipped off to the slave markets in Egypt. Soon the supply exceeded the demand and they were worthless, even as lowly slaves.

For almost 2000 years the sons of Abraham, Isaac, and Jacob have wandered around the earth with no country of their own, in constant fear of persecution and death. I am sure that they have asked the question millions of times, why all this evil for us? The true Christian has looked on with amazement and compassion, while the Jew has become a phenomenon to the world. It is no wonder that Moses wrote concerning their sufferings and punishment, "They (i.e., the judgments) shall be upon you for a sign [of warning to other nations] and for a wonder, and upon your descendants for ever" (Deuteronomy 28:46 Amplified).

Israel's history of misery which has exactly fulfilled prophetic warnings should be a sign to the whole world —a sign which among other things should teach that God means what He says, and says what He means.

Israel Reborn

The same prophets who predicted the world-wide exile and persecution of the Jews also predicted their restoration as a nation. It is surprising that many could not see the obvious: since the first part of these prophecies came true we should have anticipated that the second part would come true, also. This restoration was to come about in the general time of the climactic seven-year countdown and its finale—the personal appearance of the Messiah to deliver the new state from destruction.

Set the Record Straight

Right here a careful distinction must be made between "the physical restoration" to the land of Palestine as a nation, which clearly occurs shortly before the Messiah's coming and the "spiritual restoration" of all Jews who have believed in the Messiah just after His return to this earth.

The "physical restoration" is accomplished by unbelieving Jews through their human effort. As a matter of fact, the great catastrophic events which are to happen to this nation during "the tribulation" are primarily designed to shock the people into believing in their true Messiah (Ezekiel 38; 39).

The Scoffers

For many years prior to 1948 some Christian scholars denied the possibility of accepting the prophecies concerning the restoration of Israel as a nation in Palestine. As a matter of fact, many Bible teachers taught that all prophecy relating to Israel's future was fulfilled in Israel's past. Others taught that the promises made to Israel must be applied to the Church (since Israel rejected her Messiah). Some theologians of the liberal school still insist that prophecy has no literal meaning for today and that it cannot be taken seriously. It is difficult to understand this view if one carefully weighs the case of Israel's rebirth as a nation.

Truth From Dusty Books

There has been down through history a group of men who diligently studied the prophetic content of the Bible and took it both seriously and literally. This writer searched through many commentaries on the subject dating back to A.D. 1611 and found that many scholars clearly understood that the Jews would return to Palestine and re-establish their nation before the Messiah would come. These men held this position in spite of mocking and ridicule on the part of the majority of Christendom.

The certainty of Israel's physical rebirth as a nation and restoration to Palestine was seen by Dr. John Cumming in 1864. Thumb through his fascinating old book, over a hundred years old, and you'll read: "How comes it to pass that as a nation they have been dispersed over every land, yet insulated, separated, and alone amid the nations? The predictions of their restoration are in words as definite only not yet fulfilled. As a nation they were cut off and dispersed, and it is *as a nation that they shall be gathered and restored.*

"But one closing act in this great dramatic history of an extraordinary people is yet wanting to complete the whole. Their restoration is predicted and demanded. Who will stretch out his hand to move the scene and call forth the actors." [3]

Was this man a prophet or a student of the prophets of God?

The fact that the Jews had to be restored as a nation before Christ could return was seen by James Grant, an English Bible scholar writing in 1866.

"The personal coming of Christ, to establish His millennial reign on earth, will not take place *until the Jews are restored to their own land,* and the enemies of Christ and the Jews have gathered together their armies from all parts of the world, and have commenced the siege of Jerusalem . . . now the return of the Jews to the Holy Land, and the mustering and marshalling of these mighty armies, with a view to capturing Jerusalem, must require a *considerable time yet.*" [4]

(This was written eighty-two years before Israel was made a nation.)

Increase Mather, a famous minister in the early colonies of America, wrote a book published in 1669 entitled *The Mystery of Israel's Salvation.* In this book he developed many of the crucial prophecies of Israel's restoration. He, too, showed that the Jews would return to Palestine and become a nation before their spiritual conversion and the return of the Messiah, Jesus Christ.

The great contribution of these men who stood against the prevailing religious opinion of their day is obvious. They prove that these prophetic passages are clear and could be understood if taken literally. A hundred or more years ago the prospect of the nation of Israel seemed impossible. Their faith in these passages in the Bible has been verified before our eyes!

These men used what may be called the golden rule of interpretation which the Biblical record of fulfilled prophecy indicates is correct.

"When the plain sense of Scripture makes common sense, seek no other sense; therefore, take every word at its primary, ordinary, usual, literal meaning unless the facts of the immediate context, studied in the light of related passages and axiomatic and fundamental truths, indicate clearly otherwise." [5]

This is the method which this writer has diligently sought to follow.

Three Important Events

To be specific about Israel's great significance as a sign of the time, there are three things that were to happen. First, the Jewish nation would be reborn in the land of Palestine. Secondly, the Jews would repossess old Jerusalem and the sacred sites. Thirdly, they would rebuild their ancient temple of worship upon its historic site.

The Nation Born in a Day

Some 2600 years ago Ezekiel showed that the Jewish nation would be reborn after a long world-wide dispersion, but before the coming of the Messiah to judge a great enemy who would rise up against the new nation. Here is Ezekiel—speaking prophetically to this great enemy of the revived state—"After many days you shall be visited *and* mustered for service; in the latter years you shall go against the land that is restored from the ravages of the sword, where people are gathered out of many nations upon the mountains of Israel, which had been a continual waste; but its *people* are brought forth out of the nations . . ." (Ezekiel 38:8 Amplified).

The Time of This Prophecy's Fulfillment

There are several clues in Ezekiel's words which make it possible to pinpoint the time of this restoration.

Clue one: "The latter years" is the first. Dr. Kac, a Jewish medical doctor and noted Bible scholar, sums it up this way: "The phrase 'latter days' always refers in the Old Testament to the time of Israel's final and complete national restoration and spiritual redemption." [6]

Clue two: This restoration is clearly after a long-term desolation of the land of Israel. Note the following statements in the context: "the land that is restored from the ravages of the sword . . ." And, ". . . the mountains of Israel, which had been a continual waste. . . ."

Clue three: It is also a time when the Jewish people are being returned from exile "out of many nations." Ezekiel 37 is part of this context and details the miracle of the physical restoration of the Jews to their own land, then afterward their spiritual conversion. This was predicted to occur at a time when the world would be saying, "Behold, they say, Our bones are dried up, and our hope is lost; we are completely cut off" (Ezekiel 37:11 Amplified).

Clue four: The crux of the case is that this physical restoration to the land is directly associated with triggering the hostility which brings about a great judgment upon all nations and the Messiah's return to set up God's Kingdom. In other words, it is the presence of this reborn nation of Israel, flourishing in prosperity, that excites a great enemy from the uttermost north of Palestine to launch an attack upon them which sets off the last war of the world. This war is to be ended with such a display of divine intervention that a great many of the surviving Gentiles and Jews put their whole trust in the true Messiah, Jesus Christ.

It cannot be emphasized enough. This restoration would take place after a world-wide dispersion and long-term desolation of the land of Israel. However, it would occur shortly before the events which will culminate with the personal, visible return of the Messiah, Jesus Christ, to set up an everlasting Kingdom and bring about the spiritual conversion of Israel.

Jesus the Prophet

Jesus Christ also pinpointed the general time of His return when His disciples asked Him two important questions. "What will be the sign of your coming?" they wanted to know. And "What will be the sign of the end of the age?"

The "coming" referred to in the question above is commonly referred to as the second advent of Christ. It was only natural that they wanted to know what signs would indicate His return to set up God's promised Kingdom.

In answer Jesus gave many general signs involving world conditions which he called "birth pangs." He said that these signs, such as religious apostasy, wars, national revolutions, earthquakes, famines, etc., would increase in frequency and intensity just like birth pangs before a child is born.

One of the great signs He predicted, however, is often overlooked. He speaks of the Jewish people being in the land of Palestine as a nation at the time of His return. He speaks of "those who are in Judea" fleeing to the mountains to escape the great battles that immediately precede His return (Matthew 24:16).

Another statement of Jesus demands a national existence with even their ancient worship restored. "Pray that your flight may not be . . . on a Sabbath" (Matthew 24:20). This indicates that the ancient traditions regarding travel on the Sabbath would be in force again, thus hindering a rapid escape from the predicted invasion.

Even the Temple has to be rebuilt according to the sign given in Matthew 24:15. (More will be said about this shortly.)

Jesus' predictions regarding the nation restored to the land are extremely significant when we recall that He predicted a world-wide dispersion and complete destruction of the nation which would begin with the generation which crucified Him (Luke 21:22, 23; Matthew 23:36).

Yet when Jesus looks into the future and describes the conditions which would prevail at His coming, He puts the Jews back in the land as a nation.

It is in this context that Jesus predicts an extremely important time clue. He says: "Now learn the parable from the fig tree: when its branch has already become tender, and puts forth its leaves, you know that summer is near; even so you too, when you see all these things, recognize that He is near, right at the door" (Matthew 24:32, 33 NASB).

Perfect Parable

When the signs just given begin to multiply and increase in scope it's similar to the certainty of leaves coming on the fig tree. But the most important sign in Matthew has to be the restoration of the Jews to the land in the rebirth of Israel. Even the figure of speech "fig tree" has been a historic symbol of national Israel. When the Jewish people, after nearly 2000 years of exile, under relentless persecution, became a nation again on 14 May 1948 the "fig tree" put forth its first leaves.

Jesus said that this would indicate that He was "at the door," ready to return. Then He said, "Truly I say to you, *this generation* will not pass away until all these things take place" (Matthew 24:34 NASB).

What generation? Obviously, in context, the generation that would see the signs—chief among them the rebirth of Israel. A generation in the Bible is something like forty years. If this is a correct deduction, then within forty years or so of 1948, all these things could take place. Many scholars who have studied Bible prophecy all their lives believe that this is so.

The Repossession of Jerusalem

Another important event that had to take place before the stage would be fully set for the "seven-year countdown" was the repossession of ancient Jerusalem. Much of what is to happen to the Jewish people at the return of the Messiah is to occur in the vicinity of the ancient city.

Zechariah some 2500 years ago predicted the great invasion against the Jewish people who would dwell near ancient Jerusalem at the time of Messiah's second coming. Chapters 12 through 14 of Zechariah graphically describe the events in sequence.

Here is an outline of these crucial chapters:

1. The siege of Jerusalem by all nations (12:1–3).
2. A description of the battle in and around Jerusalem (12:4–9).
3. The personal revelation of Jesus Christ as Messiah to a remnant of Jews in Jerusalem (12:10).
4. The repentance and faith which occurs at this personal revelation (12:11–14).
5. The opening of the fountain of forgiveness to repentant Israel (13:1).
6. The triumphant return of the Messiah (14:1–21).

It is clear in these chapters that the Jews would have to be dwelling in and have possession of the ancient city of Jerusalem at the time of the Messiah's triumphant advent.

Jesus Christ also predicted this situation in His last great public message before His arrest. He warned the Jews who would be living in Judea to look for "the abomination of desolation," which was spoken of by Daniel, the prophet, standing in the "holy place" (Matthew 24:15).

"The abomination of desolation" has a technical Jewish meaning which is to desecrate the Temple by bringing a Gentile or an unholy thing into the holy place (a consecrated compartment where only an authorized priest is to enter). An "abomination of desolation" happened once before in Jewish history when in 165 B.C. an invading king named Antiochus Epiphanes slaughtered a pig in the holy place.

The point is this, in order for there to be a Temple, there would have to be a repossession of the Temple site in ancient Jerusalem.

In March and April of 1967 I was lecturing on this subject at many college campuses on the West coast. I said that if this was the time that I thought it was, then somehow the Jews were going to have to repossess old Jerusalem. Many chuckled about that statement.

Then came the war of June, 1967—the phenomenal Israeli six-day blitz. I was personally puzzled as to the significance of it all until the third day of fighting when Moshe Dayan, the ingenious Israeli general, marched to the wailing wall, the last remnant of the Old Temple, and said, "We have returned to our holiest of holy places, never to leave her again."

Needless to say, I received quite a few phone calls after that. Again, against incredible odds, the Jews had unwittingly further set up the stage for their final hour of trial and conversion.

The Third Temple

There remains but one more event to completely set the stage for Israel's part in the last great act of her historical drama. This is to rebuild the ancient Temple of worship upon its old site. There is only one place that this Temple can be built, according to the law of Moses. This is upon Mount Moriah. It is there that two previous Temples were built: the first was built by Solomon 3000 years ago. The second was built by the returning Babylonian exiles 2400 years ago. This one was completely refurbished by Herod the Great later on in an effort to win the favor and acceptance of the Jews. The second Temple was totally destroyed by Titus and the Roman Legions in A.D. 70.

There is one major problem barring the construction of a third Temple. That obstacle is the second holiest place of the Moslem faith, the Dome of the Rock. This is believed to be built squarely in the middle of the old temple site.

Obstacle or no obstacle, it is certain that the Temple will be rebuilt. Prophecy demands it.

Jesus Christ predicted an event which would trigger a time of unparalleled catastrophe for the Jewish nation shortly before His second coming. This "abomination of desolation" or desecration of the inner sanctum of the Temple would occur at the midway point of God's last seven years of dealing with the Jewish people before setting up the long-awaited Kingdom of God (Daniel 9:27).

Daniel's prediction also indicates that a prince would rise up from among the people who destroyed the second Temple (who were the Romans in A.D. 70) and that he "would make a firm covenant" with the Jewish people. This treaty would guarantee the religious freedom to reinstitute the old "sacrifices and oblations" of the law of Moses.

This "prince" must be from a revived form of the ancient Roman Empire. (More about this in a later chapter.)

The apostle Paul predicts the activities of this Roman prince in great detail and gives us insight into the act that is called "the abomination of desolation." Paul speaks of this person as one who "opposes and exalts himself above every so-called god or object of worship, so that he takes his seat in the temple of God, displaying himself as being God" (II Thessalonians 2:4 NASB). By this act, the Roman prince, who is also called "the Lawless One" and "the Antichrist," breaks his covenant with the Jewish people and causes the Jewish temple worship, according to the law of Moses, to cease (Daniel 9:27).

Tie It All Together

The main points are these: first, there will be a reinstitution of the Jewish worship according to the law of Moses with sacrifices and oblations in the general time of Christ's return; secondly, there is to be a desecration of the Jewish Temple in the time immediately preceding Christ's return.

We must conclude that a third Temple will be rebuilt upon its ancient site in old Jerusalem.

If this is the time that this writer believes it is, there will soon begin the construction of this Temple. Are there any evidences of such intentions in Israel?

In a fascinating article written shortly after the recapture of old Jerusalem, a reporter interviewed a famous Israeli historian, Israel Eldad. In answer to the question, "Do your people intend to rebuild the Temple?" Eldad said, "From the time that King David first conquered

Jerusalem until Solomon built the Temple, just one generation passed. So will it be with us."

The reporter was so startled by that answer that he asked, "What about the Dome of the Rock which now stands on the temple site?"

Eldad replied, "It is, of course, an open question. Who knows, maybe there will be an earthquake."

The hope of rebuilding the Temple that is present in the hearts of devout Jews, some of whom are in powerful positions in the Israeli government, was clearly reflected here.

With the Jewish nation reborn in the land of Palestine, ancient Jerusalem once again under total Jewish control for the first time in 2600 years, and talk of rebuilding the great Temple, the most important prophetic sign of Jesus Christ's soon coming is before us. This has now set the stage for the other predicted signs to develop in history. It is like the key piece of a jigsaw puzzle being found and then having the many adjacent pieces rapidly fall into place.

For all those who trust in Jesus Christ, it is a time of electrifying excitement.

The next war will not be with the Arabs, but with the Russians.

GENERAL MOSHE DAYAN, 1968

. . . and come from your place out of the uttermost parts of the north, you and many peoples with you, all of them riding on horses, a great host, a mighty army; you will come up against my people Israel, like a cloud covering the land. In the latter days I will bring you against my land, that the nations may know me, when through you, O Gog, I vindicate my holiness before their eyes.

EZEKIEL 38:15–16, 650 B.C.

5

RUSSIA IS A GOG

The new State of Israel will be plagued by a certain pattern of events which has been clearly forecasted.

Shortly after the restoration of the Jews in the land of Israel, an incredible enemy will arise to its "uttermost north." This enemy will be composed of one great nation which will gather around it a number of allies. It is this "Northern Confederacy" that is destined to plunge the world into its final great war which Christ will return to end.

When I was a teenager watching the end of World War II, facing the continued fear of another war, I wondered then how it would all end. I once heard a radio program with a minister saying that the Bible indicated that the last war of the world would be fought between nations symbolized by an eagle and a bear. That was interesting to me, but he didn't back up his remarks with any definite proof. Even though I wasn't religious or interested in the Bible, I still spent many hours in bull sessions about this subject with other men who were as irreligious as myself. Little did I realize at that time how definite the Bible is about who the nations will be that play the major roles in the last drama. There is certainly more revealed than the vague symbols of an eagle and a bear.

There are three major prophecies on this northern sphere of political power which are to be found in Ezekiel 38; 39; Daniel 11:40–45; and Joel 2:20. It is of paramount importance to identify the time to which these prophecies apply, who the leading nation of the confederacy is, and who the allies are.

Then we shall see what this Northern Confederacy will do and how it will end.

What Time Is It?

There are several clues in Ezekiel's prophecy which establish the time to which it applies.

First, several times in the prophecy it is ascribed to "the latter years" (Ezekiel 38:8) and "the latter days," which have been previously noted (Ezekiel 38:16). These are definite terms which denote the time just preceding and including the events which will be climaxed by the second advent of Jesus Christ, who will come this time as the "reigning Messiah" to set up God's promised Kingdom.

Second, this prediction is found in a context with a definite chronological sequence of events.

Ezekiel 36 and 37 speak of the final restoration of the Jews to the land of Palestine, a restoration from which they will never be scattered again. This restoration has two distinctions which show that it couldn't be speaking of the time when the Jews returned from the Babylonian exile.

The first distinction is that they are to return from a long world-wide dispersion. (The Babylonian dispersion wasn't very long nor was it world-wide.) The second distinction is that this restoration is immediately prior to and connected with the period of tribulation. This period brings about a great spiritual rebirth of the nation and the return of Jesus the Messiah to rescue them from their enemies.

Ezekiel speaks of the physical restoration of the nation when he says: "But you, O mountains of Israel, shall shoot forth your branches and yield your fruit to My people Israel; for they are soon to come home" (Ezekiel 36:8 Amplified).

And again from Ezekiel, "For I will take you from among the nations, and gather you out of all countries, and bring you into your own land" (Ezekiel 36:24 Amplified).

Ezekiel then foretells the spiritual regeneration of the people at some point *after* they are restored as a nation when he says, "Then will I sprinkle clean water upon you, and you shall be clean from all your uncleanness, and from all your idols will I cleanse you. A new heart will I give you, and a new spirit will I put within you: and I will take away the stony heart out of your flesh and give you a heart of flesh. And I will put My Spirit within you and cause you to walk in My statutes . . ." (Ezekiel 36:25–27 Amplified).

The parable of Ezekiel 37 describes these same events in this sequence: first, the physical restoration as a nation in the land and then the spiritual rebirth of the people. Ezekiel explains the prophetic vision, indicating the dry bones as "the whole house of Israel" hopelessly scattered throughout the nations of the world (Ezekiel 37:11). The bones coming together and sinews and flesh being put upon them is explained as meaning the regathering of the people into a physical restoration of a national existence in Palestine. Isn't it fascinating how graphic this physical analogy is?

Ezekiel's vision, however, goes beyond the purely physical. It says ". . . but there was no breath or spirit in them" (Ezekiel 37:8 Amplified). This indicates that the real spiritual life would come with the rebirth of the people after the restoration.

This restoration and spiritual rebirth of the nation is to be the beginning of the everlasting kingdom which the Messiah is promised to bring. Ezekiel says, "I will make a covenant of peace with them; it shall be an everlasting covenant with them . . . I will set My sanctuary in the midst of them for evermore" (Ezekiel 37:26 Amplified).

Study Ezekiel 38 and 39. The most significant part of this chain of events is established here. These chapters indicate with certainty that after the physical restoration of the nation, but before the spiritual rebirth, the great northern enemy will invade Israel (Ezekiel 38:8, 16). Then God will supernaturally judge the northern invaders, and this is the very act which will impel the Israeli people to know and believe in their true Messiah, Jesus Christ (Ezekiel 39:6–8).

Zechariah beautifully described this scene when he quotes God as saying, "And I will pour out upon the house of David and the inhabitants of Jerusalem a spirit of compassion and supplication, so that, when they look on him whom they have pierced, they shall mourn for him, as one mourns for an only child . . ." (Zechariah 12:10).

Ezekiel speaks in chapters 40 through 48 of a new worship pattern which will be established after the Messiah, Jesus Christ, comes to reign on earth over the Kingdom of God.

Since the restoration of Israel as a nation in 1948, we have lived in the most significant period of prophetic history. We are living in the times which Ezekiel predicted in chapters 38 and 39.

In 1854 a scholar named Chamberlain summed up the crux of what has just been said. In commenting on Ezekiel 38, he observed, "From all which I should infer, the coming restoration of Israel will at first be gradual and pacific; a restoration permitted, if not assisted and encouraged or protected. They will return to occupy the whole land, both cities and villages; they will be settled there, become prosperous and increasing in wealth, before this great confederacy of northern people will be formed against them."[1]

Consider that Chamberlain wrote this over one hundred years ago—long before Israel was a nation "assisted and encouraged" by other countries.

Who Is the Northern Commander?

For centuries, long before the current events could have influenced the interpreter's ideas, men have recognized that Ezekiel's prophecy about the northern commander referred to Russia.

Dr. John Cumming, writing in 1864, said, "This king of the North I conceive to be the autocrat of Russia . . . that Russia occupies a place, and a very momentous place, in the prophetic word has been admitted by almost all expositors."[2]

What's the Evidence?

Ezekiel describes this northern commander as "Gog, of the land of Magog, the chief prince (or ruler) of Rosh, of Meshech and Tubal" (Ezekiel 38:2 Amplified). This gives the ethnic background of this commander and his people.

In other words, the prophet gives the family tree of this northern commander so that we can trace the migrations of these tribes to the modern nation that we know.

Gog is the symbolic name of the nation's leader and Magog is his land. He is also the prince of the ancient people who were called Rosh, Meshech, and Tubal.

In the Biblical chapter commonly called the "Table of Nations" by scholars these names are mentioned. (See Genesis 10.) They are described as the grandsons of Noah through his son Japheth, with the exception of Rosh (Genesis 10:1, 2). Magog is the second son; Tubal is the fifth son; and Meshech is the sixth son.

You must be all excited about Magog, Meshech, and Tubal by this time! You are probably saying, "What in the world do these crusty relics of fiction have to do with Russia?" Let this writer assure you, these names are not fiction, but they have turned up in many archaeological discoveries in very early accounts of ancient history. One reason for this is that the families of these forefathers adopted their names as "tribal names." The family descended from Magog became known as the tribe of Magog, etc.

Dead Men Do Tell Tales!

It is necessary on the next few pages to establish some documentation from ancient history. Some people find this subject "a little dull," to say the least. If this is your case, you may wish to skim over the high points. For others, it will prove to be rewarding to check carefully the grounds upon which the historical case is built.

Herodotus, the fifth century B.C. Greek philosopher, is quoted as mentioning Meshech and Tubal. He identified them with a people named the Samaritans and Muschovites who lived at that time in the ancient province of Pontus in northern Asia Minor.[3]

Josephus, a Jewish historian of the first century, says that the people of his day known as the Moschevi and Thobelites were founded by Meshech and Tubal respectively. He said, ". . . Magog is called the Scythians by the Greeks." He continued by saying that these people lived in the northern regions above the Caucasus mountains.[4]

Pliny, a noted Roman writer of early Christian times, said, "Hierapolis, taken by the Scythians, was afterward called Magog." [5] In this he shows that the dreaded barbaric people called the Scythians were identified with their ancient tribal name. Any good history book of ancient times traces the Scythians to be a principal part of the people who make up modern Russia.

Wilhelm Gesenius, a great Hebrew scholar of the early nineteenth century, discusses these words in his unsurpassed Hebrew Lexicon. "Meshech," he says, "was founder of the Moschi, a barbarous people, who dwelt in the Moschian mountains." [6]

This scholar went on to say that the Greek name, "Moschi," derived from the Hebrew name Meshech is the source of the name for *the city of Moscow*. In discussing Tubal he said, "Tubal is the son of Rapheth, founder of the Tibereni, a people dwelling on the Black Sea to the west of the Moschi."

Gesenius concludes by saying that these people undoubtedly make up the modern Russian people.

There is one more name to consider in this line of evidence. It is the Hebrew word, "Rosh," translated "chief" in Ezekiel 38:2, 3 of the King James and Revised Standard Versions. The word literally means in Hebrew the "top" or "head" of something. According to most scholars, this word is used in the sense of a proper name, not as a descriptive noun qualifying the word "prince."

The German scholar, Dr. Keil, says after a careful grammatical analysis that it should be translated as a proper name, i.e., Rosh. He says, "The Byzantine and Arabic writers frequently mention a people called Rōs and Rūs, dwelling in the country of Taurus, and reckoned among the Scythian tribes." [7]

Dr. Gesenius in his Hebrew Lexicon says, ". . . Rosh was a designation for the tribes then north of the Taurus Mountains, dwelling in the neighborhood of the Volga." [8]

He concluded that in this name and tribe we have the first historical trace of the Russ or Russian nation.

In the light of the abundant evidence, it is no wonder that men long before Russia rose to its present state of power foresaw its role in history. Bishop Lowth of England was one of these men. He wrote in 1710, "Rosh, taken as a proper name, in Ezekiel signifies the inhabitants of Scythia, from whom the modern Russians derive their name." [9]

In the eighteenth and nineteenth centuries, such men as Bishop Lowth, Dr. Cumming, and Rev. Chamberlain, were ridiculed by many of their contemporaries. After all, who could have imagined then what we now see in modern communist Russia—a country founded upon atheism?

Where Is the Uttermost North?

The final evidence for identifying this northern commander lies in its geographical location from Israel.

Ezekiel puts great stress on this by saying three times that this great enemy of Israel would come from their "uttermost north." It is mentioned in 38:6 and 15, and 39:2. The King James Version doesn't translate this accurately, but the Revised Standard and Amplified Versions of the Bible do. The Hebrew word that qualifies "north" means either "uttermost" or "extreme."

You need only to take a globe to verify this exact geographical fix. There is only one nation to the "uttermost north" of Israel—the U.S.S.R.

"Thus says the Lord God: Are you he of whom I

have spoken in olden times by My servants the prophets of Israel, who prophesied in those days for years that I would bring you, *Gog,* against them?" (Ezekiel 38:17 Amplified). The answer to this challenging question thrown down by God through Ezekiel centuries ago is now rather obvious, wouldn't you say?

General Dayan's statement that "The next war will not be with the Arabs but with the Russians" has a considerably deeper significance, doesn't it?

Final Exam

Just think for a moment how incredible a thing we are considering here. How could Ezekiel 2600 years ago have forecast so accurately the rise of Russia to its current military might and its direct and obvious designs upon the Middle East, not to mention the fact that it is now an implacable enemy of the new state of Israel? How could men like Chamberlain and Cummings, for that matter, one hundred years ago have so clearly seen the future rise of Russia to its present world-threatening position?

The answer is again, it seems to this writer, obvious. Ezekiel once again passes "the test of a prophet." He was guided by the Spirit of the living God. In the apostle Peter's final letter, written as he faced certain and imminent death, he stated the source of the prophets' wisdom and insight. Peter first states where prophecy did not originate; "But know this first of all, that no prophecy of Scripture is a matter of one's own interpretation" (II Peter 1:20 NASB). In other words, the prophets did not dream up their own interpretation of life and history.

Then Peter declares where prophecy did originate, ". . . for no prophecy was ever made by an act of human will, but men moved by the Holy Spirit spoke from God" (II Peter 1:21 NASB).

When a man knows that he is about to die, he usually gets around to saying the things he considers to be most important. Peter considered the certainty and relevance of the prophetic word to be the most important thing. He even warned that in "the latter times" men posing as religious leaders would rise from within the Church and deny, even ridicule, the prophetic word (II Peter 2:1–3; 3:1–18).

If you pass this book around to many ministers you'll find how true this prediction has become.

Who Are the Allies?

Ezekiel partially catalogs the ancient names of the peoples and nations who would be confederates of Russia in 38:5 and 6.

Persia

All authorities agree on who Persia is today. It is modern Iran. This is significant because it is being wooed to join the United Arab Republic in its hostility against Israel. The Russians are at this moment seeking to gain footholds in Iran by various overtures of aid. In order to mount the large-scale invasion predicted by Ezekiel, Russia would need Iran as an ally. It would be much more difficult to move a large land army across the Caucasus Mountains that border Turkey, than the Elburz Mountains that border Iran. Iran's general terrain is also much easier to cross than Turkey's. Transportation, however, will be needed through both countries.

Watch the actions of Iran in relation to Russia and the United Arab Republic. This writer believes that significant things will soon be happening there.

Ethiopia or Cush (Black African Nations)

Ethiopia is a translation of the Hebrew word, *Cush.* Cush was the first son of Ham, one of the sons of Noah.

Moses mentions "the land of Cush" as originally being adjacent to an area near the Tigris and Euphrates rivers (Genesis 2:13).

After examining many authorities on the subject, the writer discovered once again why Dr. Gesenius is recognized as one of the great scholars of history. Gesenius summarized all of the evidence as follows: (1) The Cushites were black men. (2) They migrated first to the Arabian peninsula and then across the Red Sea to the area south of Egypt. (3) All the black people of Africa are descended from Cush.

Gesenius observes, "Indeed all the nations sprung from Cush and enumerated in Genesis 10:7, are to be sought in Africa."[10]

Cush is translated "Ethiopia" twenty-one times in the King James Version, which is somewhat misleading. It is certain that the ancient Ethiopians (modern Abyssinia) are made up of Cushites, but they do not represent all of them, according to history.

The sobering conclusion is this: many of the African nations will be united and allied with the Russians in the invasion of Israel. This is in accord with Daniel's graphic description of this invasion (Daniel 11:36–45). The Russian force is called "the King of the North" and the sphere of power which the African (Cush) force will be a part of is called "the King of the South."

One of the most active areas of evangelism for the Communist "gospel" is in Africa. As we see further developments in this area in the future, we realize that it will become converted to Communism.

Libya or Put (*Arabic African Nations*)

Libya is the translation of the original Hebrew word, *Put*. We have the same problem pinpointing these people as with Cush. Put was the third son of Ham (Genesis 10:6). The descendants of Put migrated to the land west of Egypt and became the source of the North African Arab nations, such as Libya, Algeria, Tunisia, and Morocco. The first settlement of Put was called Libya by the ancient historians, Josephus and Pliny.[11] The Greek translation of the Hebrew Old Testament, called the Septuagint, translates Put as Libya in about 165 B.C.

The conclusion is that Russia's ally, Put, certainly included more than what is now called Libya. Once again there are current events to show the beginning of this alliance.

The territory of Northern Africa is becoming solidly pro-Soviet.[12] Algeria appears to be already Communist and allied with Russia.

As we watch this area in the next few years we shall see indications that it is destined to join the southern sphere of power which will attack Israel along with the "King of the North."

Gomer and All Its Hordes (Iron Curtain Countries)

Gomer was the eldest son of Japheth, and the father of Ashkenaz, Riphath, and Togarmah. These people make up an extremely important part of the future Russian invasion force.

Dr. Young, citing the best of the most recent archaeological finds, says of Gomer and his hordes, "They settled on the north of the Black Sea, and then spread themselves southward and westward to the extremities of Europe."[13]

Gesenius speaks of part of Gomer's "hordes" as being Ashkenaz . . . "the proper name of a region and a nation in northern Asia, sprung from the Cimmerians who are the ancient people of Gomer. The modern Jews understand it to be Germany, and call that country by this Hebrew name. . . ."[14]

Josephus called the sons of Ashkenaz, "The Rheginians" and a map of the ancient Roman Empire places them in the area of modern Poland, Czechoslovakia, and East Germany to the banks of the Danube River. The modern Jewish Talmud confirms the same geographical picture.

The conclusion is that Gomer and its hordes are a part of the vast area of modern Eastern Europe which is totally behind the Iron Curtain. This includes East Germany and the Slovak countries.

Togarmah and All Its Hordes (Southern Russia and the Cossacks)

In Ezekiel 38:6 "the house of Togarmah, and all its hordes" are specifically pointed out as being from "the uttermost north." Gesenius says that "they are a northern nation and country sprung from Gomer abounding in horses and mules." Some of the sons of Togarmah founded Armenia, according to their own claim today, Gesenius continued.

Dr. Bauman traces evidence of some of the sons of Togarmah to the Turkoman tribes of Central Asia. This would explain the statement, ". . . of the uttermost north, and all its hordes."

The conclusion is that Togarmah is part of modern Southern Russia and is probably the origin of the Cossacks and other people of the Eastern part of Russia. It is interesting to note that the Cossacks have always loved horses and have been recognized as producing the finest army of cavalry in the world. Today they are reported to have several divisions of cavalry. It is believed by some military men that cavalry will actually be used in the invasion of the Middle East just as Ezekiel and other prophets literally predicted. During the Korean War the Red Chinese proved that in rugged mountainous terrain, horses are still the fastest means of moving a large attacking force into battle zones.

Isn't it a coincidence that such terrain stands between Russia and the Israeli?

Many People Are With You

Ezekiel indicates that he hasn't given a complete list of allies. Enough is given, however, to make this writer amazed by the number of people and nations which will be involved.

Gog, Take Command

Ezekiel, prophetically addressing the Russian ruler, commands him to, ". . . be prepared; yes prepare yourself, you and all your companies that are assembled about you, and you be a guard *and* a commander for them" (Ezekiel 38:7 Amplified).

In other words, the Russian ruler is to equip his confederates with arms and to assume command.

If you have doubts about all that has been said in this chapter, isn't it a bit unnerving to note that almost all of the countries predicted as part of this great army are already armed with weapons created and manufactured in Russia?

What's Your Game, Gog?

We have seen that Russia will arm and equip a vast confederacy. This powerful group of allies will lead an attack on restored Israel. However, Russia and her confederates will be destroyed completely by an act that Israel will acknowledge as being from their God. This act will bring many in Israel to believe in their true Messiah (Ezekiel 38:15 ff.).

The attack upon the Russian confederacy and the resulting conflict will escalate into the last war of the world, involving all nations.

Then it will happen. Christ will return to prevent the annihilation of all mankind.

Our basic aim will be to destroy Israel.
NASSER OF EGYPT, MAY, 1967

They have said, Come, and let us wipe them out as a nation; let the name of Israel be in remembrance no more.
PSALM 83:4 (Amplified) prophesied about 1000 B.C.

6

SHEIK TO SHEIK

When the phone rings and it's the person we have just mentioned in conversation, inevitably we say, "What a coincidence, we were just talking about you." Or when we open the mail and find the check that we needed to pay an urgent bill, we enjoy the timely coincidence.

However, in writing this book too many pieces and events have fallen into place for us to believe they could all be "coincidence." This is why we believe what one vague religionist has called "the divine hand from somewhere" set the stage the week we began this chapter.

In the last chapter some of the confederates of a future Russian invasion force were documented from Biblical prophecy. The confederates who are relevant to this chapter are the Arabic nations. The Bible says that Egypt, the Arabic nations, and countries of black Africa will form an alliance, a sphere of power which will be called the King of the South. Allied with Russia, the King of the North, this formidable confederacy will rise up against the restored state of Israel.

So where is the coincidence?

As we were researching the current status of the leading Arabic nation, Egypt, we discovered that on the campus of a large university, less than ten minutes away, it was Arab Week. Obviously they must have planned it to coincide with our study.

With echoes of Scheherazade humming in our subconscious, we made our way. . . .

On Campus

We found an exotic mixture of West and Middle East along the shaded walks of the beautiful university. Some students were wearing head-coverings, Arab-fashion, to indicate where their allegiance was. The vision was not quite like Lawrence of Arabia galloping across endless stretches of white sand, but the effect was striking.

Tables were everywhere around the Student Union. We were deluged with printed materials concerning the rightness of the Arab cause against Israel and their determination to liberate Palestine.

In a short while we were supplied with documents which substantiated the alliance of the Arabs and the Russians, a bond both present and actual, but also a fact which was prophesied approximately 2600 years ago!

The purpose of "Arab Week" at the university was to rally support for the Palestinian Revolution. According to the literature which was being distributed, "the Palestine Revolution draws moral support from the revolutionary movements of the world. The Arab student movement has given the revolution their total support. Arab intellectuals, joined by many world-thinkers have given their support."

This "revolutionary movement" is part of the Communist movement which has supported "wars of liberation" in countries around the world.

From the standpoint of this study of alliances which make up the King of the South we saw a valuable link in the alignment of several black African nations with the Arabs in their determined plan to "liberate" Palestine from Israel. This is another confirmation of prophecy as we have seen from Old Testament prophets.

Egypt: Leading Actor

We speak of "Arab," and yet it is obvious that the real leader in the Arabic world is Egypt. This country is in a strategic spot in the prophetic landscape, which is the reason we should follow the events in the Middle East with great interest.

Egypt is located at the southern end of "the land bridge" which connects the continents of Europe, Asia and Africa. The value of this important piece of real estate, which has been established by centuries of fighting, will play an important part in events which we will show in the chapter on World War III. Conveniently for Egypt, it has an ideal location for its role of leadership in the Afro-Arab world.

"Egypt's geographical size, large population, 150,000 man military establishment, advanced industrialization, and President Gamal Abdel Nasser's military Arab nationalism make it the political, intellectual, and cultural center of the Arab world, as well as much of Africa." [1]

What about Nasser? He has become the symbol of leadership for both African and Arabic "wars of liberation from Western imperialism." Nasser's manifesto, "Philosophy of the Revolution," provides us with insight into the future direction which we can expect of the Arab and African situation.

"Nasser views the world as a stage, with Egypt as one of the principal actors. This role is three-dimensional, described in Nasser's language in terms of circles. The first of these is the Arab area with Arab unity as the main plot. Beyond this circle lies Africa, which Nasser envisages as the seat of struggle between white 'imperialists' and the indigenous Negroes for possession of its riches. Encompassing these two circles is the world of Islam, also threatened by 'imperialism.' In the past few years, Nasser appears to have expanded this conception of the third circle to include all non-Western and underdeveloped Western countries." [2]

Nasser has not swerved from his written goal to bring about a kind of "Arab Socialism." He has repeatedly said that kings, sheiks, sultans, and capitalism must all be obliterated. This has appealed to the common Arab who has been oppressed for centuries. Using the "Gospel of Materialism," plus the common bond of Arabic race identity, wedded with the Moslem religious ties, Nasser believes that he can unite the

Arabs to lead the resurrection of all underprivileged nations into a mighty third world force. He envisions himself as the one to lead the nations of Africa, black and Arab, to unity.

Somehow the aims, ambitions, and worldly directions of dictators past and present never seem to change. There has never been a benevolent dictator.

How to Make Enemies and Influence People

Nasser has fallen into a trap which has ensnared all Arab leaders. It would appear that the only way to remain a popular leader in the Arab world today is to keep the flames of hatred toward the state of Israel fanned to a fever pitch. The one who can make the most elaborate and gory promises of Israel's destruction is number one on the hit parade. Whenever an Arab leader senses his popularity waning, he whips up a propaganda program about the need to liberate Palestine, according to Middle-East observers.

It is believed by most experts on the Middle East that Nasser was trapped into the June, 1967, war, playing the game of Blind Man's Bluff. He knew that he could never remain enthroned as the leader of the United Arab Republic if other aggressive Arab leaders railed against Israel in stronger terms. It is reported that he was caught unprepared when U Thant quickly complied with his order to remove the U.N. observers from the buffer zone which separated the Arabic and Israeli armies. Once this occurred, he had no alternative but to make good his threats.

Israel saw the clear danger of a large-scale Egyptian mobilization in the Sinai peninsula and also the threat of not being able to ship through the Gulf of Aqaba. The Israelites also saw the rapid unification of all Arabic nations into a formidable force surrounding them on three sides. Israeli leaders realized that unless they seized the initiative and attacked, there would be no hope of survival.

What started out to be a bold popularity stunt on Nasser's part, designed to make Israel lose face over the blockade of the Gulf of Aqaba, ended in a fiasco.

Rather than be completely humiliated in the eyes of the world, Nasser carried the world to the brink of war.

It is this kind of fierce pride and smoldering hatred against Israel that will keep the Middle East a dangerous trouble spot. No Arab leader could hope to remain in power if he were willing to make concessions in negotiating with Israel.

In July, 1968, the headlines in a news magazine warned, "No Easing of Mideast War Danger." The report said that "The recent visit of Egypt's President Nasser to the Soviet Union—the country that arms him against Israel—turned attention to an area that never seems far from the explosion point." [3]

In December, 1968, the ambassador to the United States from Israel, Yitzhak Rabin, a key strategist in the six-day Arab-Israeli War of 1967, said that he couldn't be optimistic about peace in the near future in the Middle East.[4]

U Thant said of the Egypt-Israeli situation, "Never in the history of the United Nations' experience with peace-keeping has there been such complete and sustained disregard for a cease-fire agreed to by the parties." Mr. Thant went on to say that warfare along the Suez Canal had become so intense he might have to consider withdrawal of U.N. cease-fire observers.[5]

At the time this is being written Nasser is reported to be in poor health. Whether he continues to lead Egypt, or is replaced by some other leader or dead by the time this is published, the clearly predictable course of the Middle East will not be changed. There will be continual crises there and a great involvement of the world's major powers.

The King of the South

Current events in the Middle East have prepared the stage for Egypt's last act in the great drama which will climax with the finale, Christ's personal return to earth.

We are not attempting to read into today's happenings any events to prove some vague thesis. This is not necessary. All we need to do is know the Scriptures in their proper context and then watch with awe while men and countries, movements and nations, fulfill the roles that God's prophets said they would.

Long ago the prophet Daniel spoke of Egypt as "the king of the South." Egypt is identified as this power in chapter 11 where Daniel predicts a long span of history involving warfare between Egypt under the Ptolemaic dynasty and Syria under the Seleucid dynasty.

In Daniel 11:40 Daniel leaps over a long era of time to the events which lead up to the personal, visible appearance of Christ as God's righteous conqueror. The phrase "at the time of the end" speaks unmistakably of the beginning of the last great war of history.

Daniel gives great detail concerning the battles and movement of troops which will take place at the beginning of this war. (The war itself will be developed later in chapter 12.)

Our interest here is the revelation that Egypt will attack the revived state of Israel, which will then be under the control of a false Messiah. This man will probably be a Jew who works closely with the world dictator who will come to power in Rome. (Meet the "Future Fuehrer" in chapter 9.)

Notice what Daniel says about this attack on Israel: "And at the time of the end the king of the south shall push at and attack him" (Daniel 11:40 Amplified).

This immediately triggers another invasion of Israel by Russia who is here called "the king of the north."

The movement of Russia and its northern confederacy through "the land bridge of the Middle-East" into Egypt serves an ominous warning to Egypt. Speaking of the Russian invader, Daniel prophesies: "He shall stretch forth his hand also upon the countries: and the land of Egypt shall not escape. But he shall have power over the treasures of gold and of silver, and over all the precious things of Egypt; and the Libyans and the Ethiopians shall be at his steps" (Daniel 11:42, 43 KJV).

As we saw in the last chapter, the Hebrew words "Cush" and "Put," which are translated Ethiopia and

Libya, represent the black Africans and African Arabs, respectively. Aside from the obvious evidence of a Russian double-cross of the Egyptians, this passage also indicates that the "black African" and "Arab-African" countries will be involved with Egypt and in line for Russian conquest as well. The statement, ". . . the Libyans and Ethiopians shall be at his (the Russian invader) steps," indicates one of two things: they will be next in line for conquest, or they will submit totally to the Russian will and be assimilated into the northern confederacy.

This invasion of Cush and Put, along with Egypt, and their fall together is mentioned more specifically by the prophet Ezekiel: "And a sword shall come upon Egypt, and anguish shall be in Ethiopia (Cush), when the slain fall in Egypt, and her wealth is carried away, and her foundations are torn down.

"Ethiopia, and Put, and Lud, and all Arabia, and Libya, and the people of the land that is in league shall fall with them by the sword" (Ezekiel 30:4, 5).

This prophecy, in the first nine verses of Ezekiel 30, refers to the judgment of Egypt and her allies during the Tribulation. The phrases, ". . . the day of the Lord" and ". . . a time of doom of nations" place it in the time just prior to the second coming of Christ. For you students of the Bible, we must add that the latter part of the chapter looks at the time when Nebuchadnezzar destroyed Egypt and her allies, but its greater fulfillment is future.

Are you discovering more pieces of this stirring prophetic puzzle? The Egyptian plan to unite the Arabs and the black Africans into a "third world force" seems to be fulfilling what the prophets have said.

Grace Goes to Egypt

After viewing such a bleak picture of the future for Egypt and its confederacy of nations it may seem as though God has written them off. The truth of the matter is, however, quite the contrary.

Isaiah, reliable prophet that he is, reveals that one

of the purposes for the judgment of Egypt is to drive its people from faith in false messiahs and "religion" to faith in the one true Savior.

Isaiah warns of a terrible judgment which would fall on Egypt in the last days. He speaks of Egypt's very life source being judged: "And the waters of the Nile will be dried up, and the river will be parched and dry; and its canals will become foul, and the branches of Egypt's Nile will diminish and dry up" (Isaiah 19:5, 6).

If you think the famous Aswan Dam, which diverts the main channel of the Nile River, will help the Egyptian situation, you're mistaken. Somehow the headwaters of the Nile will be diverted and that important river will be a parched piece of real estate. Imagine the terrifying implications of this to an Egyptian!

Isaiah warns of a powerful dictator who will invade and take them over: ". . . I will give over the Egyptians into the hand of a hard master; and a fierce [merciless] king will rule over them" (Isaiah 19:4 RSV). This refers to the Antichrist of Rome who will possess Egypt after Russia is destroyed.

All of these things will happen to the Egyptians until many cry out to the true Savior, Jesus. Isaiah says, ". . . when they cry to the Lord because of oppressors he will send them a savior, and will defend and deliver them" (Isaiah 19:20).

What a great demonstration of God's loving heart! Often men won't see their need of God until He so shakes up their world that they are helpless to cope with life without Him. It's only then that they turn to trust in God's provision for their shortcomings. Then they discover that Jesus Christ has so paid the penalty for their sins that God can offer a totally free gift of forgiveness and accept them into His eternal family.

A Lesson From Egypt

As you read this book you may have reached the point where you recognize your inability to live in a way that would cause God to accept you. If this is the case, you may speak to God right now and accept the gift of Christ's forgiveness. It's so simple. Ask Christ to come into your life and make your life pleasing to God by His power.

We have found the results to be certain and exciting in our own lives.

Putting It Together

We have seen how current events are fitting together simultaneously into the precise pattern of predicted events. Israel has returned to Palestine and revived the nation. Jerusalem is under Israeli control. Russia has emerged as a great northern power and is the avowed enemy of revived Israel. The Arabs are joining in a concerted effort to liberate Palestine under Egyptian leadership. The black African nations are beginning to move from sympathy toward the Arabs to an open alliance in their "liberation" cause.

It's happening. God is putting it all together. God may have His meaning for the "now generation" which will have a greater effect on mankind than anything since Genesis 1.

Will you be ready if we are to be a part of the prophetic "now generation"?

. . . the great river, the Euphrates . . . was dried up, that the way might be prepared for the kings of the east . . . And they gathered them together in the place which in Hebrew is called Har-Magedon.

REVELATION 16:12, 16 (NASB)
JOHN THE APOSTLE about A.D. 90

7

THE YELLOW PERIL

"The kings of the East" in Biblical prophecy refers to another sphere of power which was to arise in the world at the same time as the great Northern Power (Russia) and the King of the South (Egypt and the Afro-Arabic alliance).

The original Greek words translated "east" (Revelation 16:12) are literally *anatoles heliou,* which mean, "the rising of the sun." This was the ancient designation of the Oriental races and nations. John describes this vast horde of soldiers assembled at the Euphrates River as "the kings of the sun rising" and thus definitely predicts the movement of a vast Oriental army into a war in the Middle East.

The mention of the Euphrates River brings out another important strategic clue about this Eastern confederacy. This great river has figured prominently in military history throughout the ages. It was always recognized as the ancient boundary between east and west. In the late nineteenth century a scholar said of this fact: "From time immemorial the Euphrates, with its tributaries, has been a great and formidable boundary between the peoples east of it and those west of it. It runs a distance of 1800 miles, and is scarcely fordable anywhere or at any time. It is from three to twelve hundred yards wide, and from ten to thirty feet in depth; and most of the time it is still deeper and wider." [1] This clue shows that this power is Oriental, since it comes from east of the Euphrates.

The Euphrates has presented a formidable problem for the corps of engineers of many ancient armies of the past. In this future invasion, however, God Himself will see to it that the river is dried up so that a trap is set for triggering the last great war of mankind.

Another important detail involving this Oriental army is unlocked by the clue involving the Euphrates. The apostle John speaks of the release of four vicious, depraved angelic beings which have been kept bound by God at the Euphrates River (Revelation 9:14–16). Immediately after their release an incredible army emerges from the Euphrates . . . it numbers "200 million" (Revelation 9:16). The four demonic spirit-beings somehow incite this great army to invade the Middle East and apparently they are the ones who make the river dry up so that the army can quickly cross this ancient barrier of east and west.

A terrifying prophecy is made about the destiny of this Asian horde. They will wipe out a third of the earth's population (Revelation 9:18). The phenomena by which this destruction of life will take place are given: it will be by fire, smoke (or air pollution), and brimstone (or melted earth). The thought may have occurred to you that this is strikingly similar to the phenomena associated with thermonuclear warfare. In fact, many Bible expositors believe that this is an accurate first-century description of a twentieth-century thermonuclear war.

Another Coincidence?

These predictions, made near the end of the first century, concerned an Asian confederacy which would field the greatest army ever to march to battle. This army would be raised up just prior to Christ's return to the earth.

For centuries Asia has had a tradition of backwardness. Though the peoples of Asia have always been numerous in population, they lagged behind the West in education, science and technology. For hundreds of years Asia chose to remain isolated from the world; then that isolation was broken.

Japan was the first Asian nation to push out into the stream of modern science; it was the first Asian nation in centuries to set out on a course of conquest beyond the Orient. But Japan never truly got outside of the Asian boundaries.

The Japanese navy almost brought a great invasion force into the Middle East during World War II. A convoy set sail for the Red Sea to enter the African and Palestinian campaign and break the Allied resistance there. Nothing could have stopped them. They would have defeated the British who were already extended by Field Marshal Rommel's Afrika Corps.

The British navy had only a handful of ships in the Indian Ocean and they were quickly ordered to flee to Madagascar to avert suicide. Had the Japanese continued as originally planned, World War II could have had a different ending.

It was at this point that a strange thing happened. Admiral Yamamoto, for some unexplainable reason, changed the order, and had the task force turn around in the Indian Ocean and head for the West Coast of the United States. We believe this must have been divine providence.

The intention of this task force was miraculously discovered when some sailors of the U.S. Navy intercepted their radio messages and broke the code. The battle that ensued was really the turning point of the war. With a handful of B-17 bombers and a vastly outnumbered task force, the U.S. Navy turned back the Japanese at the Battle of the Coral Sea.

Japan went on to be defeated and her dreams of world conquest have been put away.

However, the churning of upheaval was taking place in the Orient.

The Dragon Is Awake

With the Communist takeover of China, the real sleeping giant of Asia was awakened. In about 1860 an astute student of prophecy, Dr. Robinson, predicted:

"Before another half century shall have rolled away in the providence of God there will be seen revolutions in the Oriental mind of which no one has even a foreboding." [2]

Dr. Cumming in 1864 also foresaw the necessity of the Orient entering the industrial age and later becoming a great scourge to the Western civilization.[3]

In the twenty odd years since the fall of China to the Communists there has been a steady relentless preparation for all-out war with the free world. Though the living conditions of the eight hundred million or more people of Red China are still basically like the nineteenth century, they have made remarkable progress in the production of weapons for war.

Concerning Red China's potential and purpose, Victor Petrov said in 1967, "China does possess all the prerequisites for being or becoming a world power. Communist China's economic growth is evident and has been on the rise. With or without Soviet help, she will be progressing toward her avowed goal of reaching the industrial level of the other major powers of the world . . . a giant, for decades half-asleep, sheepishly watching the rest of the world go by on the path of technological progress. This giant has apparently been awakened." [4]

Red China is definitely on the road to becoming a world power, but her design for the use of this stature is not peace. Within one year after the take-over of China her Communist leaders started the war in Korea. They have since fomented the war in Viet Nam and have traveled to several countries in Africa and the Middle East seeking to aggravate internal subversion and "Communist wars of liberation."

The vaunted Sino-Soviet split is over an interpretation of Communist doctrine. The Chinese insist that the world can be captured only by force of arms and violence: the Russians now believe that the free world can be captured by the relatively limited violence of internal subversion . . . while masquerading under the guise of "peaceful coexistence." It should be marked well, however, that neither have disembarked from their goal of total world conquest for Communism. This is an integral part of the Communist doctrine. Without the total destruction of the capitalist system the basic promise and goal of Communism could not be attained, that is, the changing of man's nature by the complete change of his environment. According to the Communists, as long as capitalism exists in the world it continues to infect man's environment and prevents him from being a creature that loves to work, shares equally his wealth, and loves his fellow man.

The great charge of the Red Chinese against the Russians is the most despicable word in the Communists' vocabulary—"revisionist." They believe that the Russians have "revised" the most fundamental principle of Marxist-Leninism. Lenin succinctly stated this principle. "Marxists have never forgotten that violence will be an inevitable accompaniment of the collapse of capitalism. . . ."[5]

Mao Tse-tung summarized his interpretation of this principle when he said, "Political power comes out of the barrel of a gun . . . the gun must never slip from the grasp of the Communist party."[6]

The basic difference can be described as "external invasion and takeover" versus "internal subversion and takeover."

Because of the Communist Chinese belief that the free world can only be overthrown by all-out war, they have for many years devoted approximately ten per cent of their entire military budget for developing nuclear weapons. In the February, 1969, issue of the *Bulletin of Atomic Scientists,* which was on the theme of "China's Nuclear Options," special note was made of the fantastic technological feat achieved by China's development of the H-Bomb. They went from the testing of a crude atomic bomb to the successful test firing of an H-bomb in two and one half years. This was much faster than the time taken by the other members of the world's Atomic Bomb Club.

In the same issue of the *Bulletin of Atomic Scientists,* Michael Yahuda discussed the various options which the Red Chinese have for a delivery system for H-bombs. He expressed the following opinion: "The third option —an ICBM based strategy—would be the most satisfying psychologically to the present Chinese leadership and to Mao in particular. At one fell stroke, the Chinese would have acquired the most advanced weapon . . . The American mainland would be within range as indeed would the Urals and European Russia. Current, although incomplete, evidence suggests that the Chinese are striving for an ICBM capability." [7]

Yahuda's speculation is certainly shared by the military planners. This was undoubtedly the prime motivation behind the Nixon administration's push for an anti-ballistic missile system.

Dr. David Inglis wrote in the February, 1965, *Bulletin of Atomic Scientists* regarding this threat, "Our concern should anticipate at least two decades ahead. In such a time the large human and material potential of an upsurging China constitutes a nuclear threat so vast that no effort should be spared to anticipate this threat." [8] This was written before China's successful testing of the H-bomb.

Mao may be dead before this is read, but the course of China's path of destruction will continue. The new leaders of Communist China may be more unstable than Mao.

We believe that China is the beginning of the formation of this great power called "the kings of the east" by the apostle John. We live at a time in history when it is no longer incredible to think of the Orient with an army of 200 million soldiers. In fact, a recent television documentary on Red China, called "The Voice of the Dragon," quoted the boast of the Chinese themselves that they could field a "people's army" of 200 million militiamen. In their own boast they named the same number as the Biblical prediction. Coincidence?

Furthermore, the Chinese leaders claim that even nuclear weapons cannot stop their human wave tactics. They brag about the invincibility of the unbelievable numbers of soldiers they can expend in a given campaign. This "human wave" tactic seems to be behind the strategy of that 200 million man army predicted as invading the Middle East in the battle of Armageddon. Petrov says regarding this, "The abundant population presents an unlimited source of human material for military power, which in our age of machinery and automation has not yet been found obsolete. China's armed forces are formidable in numbers . . . that day apparently is approaching when China will definitely become a member of that small but exclusive group commonly known as the Great World Powers."[9]

The sobering fact that Red China will have ICBM's capable of delivering H-bombs by 1980 at the latest, presents another grisly potential for fulfilling prophecy regarding this Oriental power. Within a decade China alone will have the capacity to destroy one-third of the world's population just as John predicted.

Summing Up

We believe that another sphere of political power is forming its predicted role in the final stages of history. Along with the revival of Israel and the return of the dispersed Jews, the rise of Russia, the formation of the Arab confederacy, China is helping to shape the Orient into its pattern of prophecy.

History seems to be headed for its climactic hour.

Veni, vidi, vici.
CAESAR, 47 B.C.

8

ROME ON THE REVIVAL ROAD

When Caesar sent his memorable dispatch, "I came, I saw, I conquered," the scribes of his day might have said, "We shall record these historic words immediately. Perhaps Latin students centuries from now will be required to memorize them."

Students who have struggled through Latin declensions and waded with Caesar through campaigns in Gaul know his famous words well. Certainly many other men have expressed themselves more eloquently. But there haven't been many men through the centuries who have had the power of Julius Caesar.

But Rome fell. And Caesar died as any mortal must. And the mighty Roman Empire of the ancient world lost its strength.

However, the prophetic Scriptures tell us that the Roman Empire will be revived shortly before the return of Christ to this earth. A new Caesar will head this empire and "Veni, vidi, vici" will leap out of the first-year Latin books and become a reality of the times.

Twenty years ago no one would have dared to believe that Rome as an empire would be put back together. And yet we are seeing significant movements of nations today which are indications that this is what is happening.

As world events develop, prophecy becomes more and more exciting. Also, the understanding of God's prophecies becomes increasingly clear as we look at the Bible and then at the current scene.

We are told in Daniel 12 how prophecy "will not be

understood until the end times, when travel and education shall be vastly increased."

We are told that "Surely the Lord God will do nothing without revealing His secret to His Servants the prophets" (Amos 3:7 Amplified).

In other words, when God is going to undertake some significant movement of history as far as His program is concerned, He will reveal it first. This writer doesn't believe that we have prophets today who are getting direct revelations from God, but we do have prophets today who are being given special insight into the prophetic word. God is opening the book of the prophets to many men. This is one reason you will find on Christian bookshelves an increasing number of books on the subject of Bible prophecy.

All Roads Lead To . . .

Where in prophecy do we find a prediction about the revival of Rome? First we shall examine the great prophet, Daniel. Chapter 7 of Daniel was written sometime in the early sixth century B.C., at a time when Babylon was still the ruling empire of the world. In the first part of this chapter the prophet Daniel was shown the successive empires which would come on the stage of history and have authority over the whole earth. A vision of these empires was also described in Daniel 2. The key to these empires is given in Daniel 2:39 where he predicts the successive empires which shall "bear rule" (literally meaning, shall have authority) over all the earth.

First of all, there was Babylon. "You (king of Babylon) are the head of gold."

Then the prophet tells us what kingdoms will rise to power after Babylon.

"And after you shall arise another kingdom [the Media-Persian], inferior to *and* earthward from you, and still a third kingdom of bronze [Greece under Alexander the Great], which shall bear rule over all the earth. And the fourth kingdom [Rome] shall be strong as iron, since iron breaks to pieces and subdues all things; and like iron which crushes, it shall break and crush all these" (Daniel 2:39, 40 Amplified).

These kingdoms would conquer everything that was worth conquering on the known earth of that time.

The Greatest Chapter in the Old Testament

The seventh chapter of Daniel, written before the coming of Jesus of Nazareth, was known by the scribes as the greatest chapter in the Old Testament. Jesus and His apostles referred to it directly or indirectly many times. Many of the predictions from Daniel's dream have been clear in their historic perspective for centuries. However, certain parts have remained obscure until recent times.

Daniel had a dream, and in this dream he saw four great beasts come up out of the sea. The first beast was like a lion, but had eagle's wings. The second beast was like a bear; the third beast was like a leopard, but had four heads. The fourth animal was "dreadful and terrible"—it had iron teeth and ten horns.

If you were an interpreter of dreams, how would you feel if you had a nightmare like Daniel's? Probably you can sympathize with the prophet when he said he was confused and disturbed. But we do not need to be confused by the outward complexity of these descriptions. Daniel had it explained to him by the angels who were the official interpreters of this vision. "These great beasts, which are four, are four kings, which shall arise out of the earth" (Daniel 7:17 KJV).

The Four Kingdoms

The first kingdom was Babylon, which became a world empire in 606 B.C. when it conquered Egypt. Nebuchadnezzar took over the Babylonian Empire upon his father's death and made it a world kingdom.

The second kingdom, which was like a bear, was the Media-Persian Empire (Daniel 8:20). The Babylonian empire was conquered by the Medes and Persians about 530 B.C. when they ingeniously built the dam in the Euphrates River.

For a time the Media-Persian Empire was great. The first two kings became believers in the God of Israel. But Daniel predicted long before the Greeks grew in power, when their future leaders were obscure Macedonian hillbillies, that they would become strong and defeat the Media-Persian Empire. And so they did. In 331 B.C. Alexander the Great conquered the Persian Empire and took it over. The third empire, according to prophecy, became a reality in history.

As predicted in Daniel 8, the Greek Empire disintegrated when the first king died prematurely. It was also predicted that four powers from within would divide the empire. And so it happened. Four generals of Alexander the Great took over the empire and divided it into four parts—they lasted until about 68 B.C. when the Romans conquered the last part of the ancient Greek Empire. It was then that Rome became the greatest world power to that date.

If you are a careful Bible student you know the common sport in the classroom today, especially in courses called "The Bible as Literature," or something similar. Teachers love to tear the Book of Daniel apart—they especially like to late-date it. Some liberal professors claim that it was written in 165 B.C., in order to discredit the supernatural element of prophecy. However, the authenticity of Daniel and its early date has been carefully defended by such scholars as Dr. Merrill F. Unger,[1] Dr. E. J. Young,[2] and Sir Robert Anderson.[3]

Focus on the Fourth Kingdom

The fourth kingdom, Rome, was not given the name of any animal, but it would be a beast unlike any other—more ferocious than all the rest. "Then I would know the truth of the fourth beast, which was diverse from all the others, exceeding dreadful, whose teeth were of iron, and his nails of brass; which devoured, brake in pieces, and stamped the residue with his feet" (Daniel 7:19 KJV).

This verse speaks of the first phase of this fourth kingdom. In phase 1 this kingdom gains world authority (as Rome did), and then disappears to emerge again just before Christ returns to establish the Kingdom of God.

Rome: Phase 2

In phase two of the fourth kingdom, Rome, the kingdom will be in the form of a ten-nation confederacy. "And of the ten horns that were in his head, and of the other which came up, and before whom three fell; even of that horn that had eyes, and a mouth that spake very great things, whose look was more stout than his fellows" (Daniel 7:20 KJV).

The meaning of these symbols will become clearer as we continue in Daniel. To the uninitiated these Biblical pictures may seem to resemble the famous Dr. Seuss animals. However, Biblical symbolism is established in historical fact.

Daniel continues with his vision: "I beheld, and the same horn made war with the saints, and prevailed against them; Until the Ancient of days came, and judgment was given to the saints of the most High; and the time came that the saints possessed the kingdom" (Daniel 7:21, 22 KJV).

The "Ancient of days" who is described here is identified in Daniel 7:13 as "one like the Son of Man" brought on the clouds of heaven—the One who will put down all human authority and establish His kingdom forever.

There is a cohesiveness to the Scriptures that is fascinating. Jesus knew the prophecies of His coming. When He was on trial before the Sanhedrin, which was the Jewish high court, He was put under oath by the high priest to tell who He really was. He was asked directly, "Are you the Son of God?" Jesus answered, "I am." He also said, "Ye shall see the Son of man sitting on the right hand of power, and coming in the clouds of heaven" (Mark 14:62–64 KJV).

Jesus was referring to this verse in Daniel 7:13 and everyone in that court of law knew what He was talking about. The high priests knew the prophets and their writings. They were furious with Jesus and proclaimed

Him a blasphemer because He was claiming to be the "Ancient of days" who was coming to set up God's kingdom on earth.

To continue with the second phase of the Roman Empire, the Scripture says that the ten horns which were just described are ten kings, or ten nations: "And the ten horns out of this kingdom are ten kings that shall arise: and another shall rise after them; and he shall be diverse from the first, and he shall subdue three kings" (Daniel 7:24 KJV).

When the Scripture says "out of" it means the ten nations (ten kings) which will come out of Rome, since Rome was the fourth kingdom. But who is "another"? This is the beast, the Antichrist.

After these ten nations arise out of the cultural inheritance of the Ancient Roman Empire, another king shall rise "diverse from the first." In other words, he will be different. He will not only be a political leader, but a religious leader. (We shall see this in the next two chapters.) When he comes to power, he will subdue three of these kings or nations. Seven of them, however, will give him authority willingly.

. . . Couldn't Put Rome Together Again

The Roman influence upon the world is so extensive that it touches Western civilization in every aspect of life. From absorbing epics like *Ben Hur* to the Roman candle we shoot on the Fourth of July, we are saturated with the glory that was Rome. However, Rome disintegrated from within; unfortunately, there is in America the same trend in moral decay that led to the downfall of Rome.

It is interesting to see in history how men have attempted to put together the old Roman Empire. Charlemagne tried to do this in A.D. 800. His "Roman Empire" included what are now the countries of France, Germany, Italy, Holland and Belgium. Charlemagne was crowned by the Pope as Emperor Charles Augustus. But his empire was not the ten-nation confederacy of the Scriptures.

Napoleon tried his strutting best to establish his own

Roman Empire. Another Pope, Pius VII, made a tedious trip across the Alps to Notre Dame cathedral in Paris to place an imperial crown on Napoleon, but the new little Caesar snatched the crown from the Pope and put it on himself. His empire was not the revived Roman Empire, either.

And then there was Hitler. Does anyone doubt that he attempted to put Rome together again? He said his Third Reich would last a thousand years. God had other plans and Hitler lost.

In spite of the vain striving of man, of the bold and infamous conquerors throughout the ages who failed in their human attempts, we are beginning to see the Ancient Roman Empire draw together, just as predicted.

If you are racing for your Rand McNally, please spare yourselves the exertion. We are not speaking of a revived Roman Empire in the physical, geographical sense, although some of these countries were part of the Ancient Roman Empire, but we are speaking of those countries which are the depository of the people, the culture, and tradition of Rome.

United We . . . Must

If the formation of the European Common Market were an isolated development in the line of Biblical prophecy, then it would have no significance for our study. However, combined with the other pieces of the prophetic puzzle which we are attempting to develop for you, it takes on immense importance.

We believe that the Common Market and the trend toward unification of Europe may well be the beginning of the ten-nation confederacy predicted by Daniel and the Book of Revelation.

What particular forces are contributing to the evolvement of this federation that the conquerors of history could not command?

First, there is the threat of Communism. One of the great motivating factors in forming this economic community and NATO was the concern over a common enemy. An article about "Mister Europe at Eighty," quotes Jean Monnet, called the father of the Common Market, as saying, "As long as Europe remains divided, it is no match for the Soviet Union. Europe must unite." [4]

The second reason for the formation of the European Common Market was the economic threat of the United States. Europeans realized they could not survive the industrial might of the United States. A charismatic personality on the present-day scene is Jean-Jacques Servan-Schreiber, newspaper editor and author of *The American Challenge,* a book which has received great acclaim throughout Europe. Some have said that the handsome Frenchman is pushing hard for a United States of Europe because he wants to become its first president. Whether this is true or not we would not venture to guess. However, it does seem that he is "furiously trying to push a U.S. of Europe," as one writer expressed it.

Servan-Schreiber is quoted as saying that "a successful response to American technology, organization, and research demands a united European effort." [5]

The third reason this writer feels Europe will form this ten-nation confederacy is that Europeans sense the basic weakness of the United States in its will to resist Communism. They seem to realize that if Europe were really at stake the U.S. would be dragging its feet in reacting against a Russian invasion. As an American it is difficult to write these words, but Europe does not feel that it can count on us in a real showdown.

A fourth factor is that according to the prophetic outlook the United States will cease being the leader of the West and will probably become in some way a part of the new European sphere of power.

Sheath your weapons, please. We realize that the United States is not mentioned in the Bible. However, it is certain that the leadership of the West must shift to Rome, in its revived form, and if the U.S. is still around at that time, it will not be the power it now is.

In spite of many who propose alternatives to a United States of Europe, and the temporary setbacks it appears to have, it seems that the trend is ever onward. An American business magazine said, "Despite its tendency to hang dangerously over cliffs, the Common Market is here to stay." [6]

A fifth factor in the trend toward the ten-nation confederacy is the realization of the great potential of a United Europe. Many men have been preaching this, not just Servan-Schreiber. A few years ago the French foreign minister said that the Common Market will carry with its network of interests and involvements across the world such weight that it will eventually become a world system.

Former Secretary of State Dean Rusk said, "Powerful forces are moving in the European community toward political integration as well. Survival and growth force the nations of Europe to forget their historic antagonisms and unite. Through the pooling of the resources and efforts a mighty new entity is growing out of the chaos left by national rivalries and world wars." [7]

A friend who lives in Germany sent us this translated statement of Dr. Walter Hallstein, who was formerly the president of the European Economic Community. Our correspondent, knowing our interest in Bible prophecy, thought there was significance in Hallstein's words. We'll let you judge.

"Three phases of the European unification are to be noted. First, the customs union, second, the economic union, third, the political union . . . what we have created on the way to uniting Europe is a mighty economic-political union of which nothing may be sacrificed for any reason. Its value exists not only in what it is, but more in what it promises to become . . . At about 1980 we may fully expect the great fusion of all economic, military, and political communities together into the United States of Europe."

Hallstein cited 1980. The timetable may be accelerating. Developments in Europe have changed so rapidly that an American news magazine had a feature headed "Europe's Dreams of Unity Revive."

One sentence in that story leaped from the page: "Should all go according to the most optimistic schedules, the Common Market could someday expand into a ten-nation economic entity whose industrial might would far surpass that of the Soviet Union." [8]

Imagine that. A "ten-nation economic entity."

Is it any wonder that men who have studied prophecy for many years believe that the basic beginning of the unification of Europe has begun?

What Else Is New?

At the time that this Roman Empire will begin to be revived there will also be a revival of mystery Babylon. If this sounds rather spooky, bring your head out from under the skeptical covers and examine with us in a later chapter the Biblical basis and the current applications.

Heading the revived Roman Empire will be a man of such magnetism, such power, and such influence, that he will for a time be the greatest dictator the world has ever known. He will be the completely godless, diabolically evil "future fuehrer."

The spirit that I have seen may be a devil; and the devil hath power to assume a pleasing shape.

HAMLET

9

THE FUTURE FUEHRER

A dictator? Who is a dictator? What makes a dictator? A dictator is a person with absolute authority, a person who has power over people. Does he appear suddenly on the scene and say, "Stop this outmoded democratic process, I am your leader now"? That isn't the way it happens.

A dictator does not thrust his rule upon people from the top down, without provocation. His tyranny is the end result of chaos in the society that results in his rise to despotic power. The dictionary describes him as a person who seizes the authority over a nation as the result of an emergency.

The chicken and egg principle does not apply here. A troubled society produces the atmosphere for the rise of a dictator, it is not the dictator who initiates the conditions which account for his rise. However, once established as the "Big Cheese," history has shown that the dictator cannot find lasting solutions to the problems.

The power-mad leader of the Third Reich, Adolf Hitler, would not have achieved his terrifying control over the lives of millions if the times had not been ripe for him. In the 1930's the German people were in despair. Economic depression was shaking the foundations of industry. Millions were out of work. Small business enterprises were collapsing. It was a desperate time, with desperate people looking for a way out. Hitler, with his evil genius, knew the mood of the German people. He saw the opportunity to pursue his Putsch in precisely the era that history demanded.

Hitler saw himself as a hero, a savior, a strong man needed by weak underlings—a "great leader" who could guide the Germans to heights of glory, according to his standards. He believed he was above the moral standards of ordinary man. Consequently, he surrounded himself with shady characters of every type. As long as they were useful to him this motley bunch of criminals and sadists were part of the ruling strength of Hitler's inner circle.

When in Rome

Hitler took the name for his empire from the all-powerful First Reich, which was the Holy Roman Empire. It was in Rome that the Caesars introduced universal Caesar worship. Their dictatorial powers, also, were absolute.

A Scottish Bible scholar has written: "The extraordinary fact is that emperor worship was not imposed on the Roman Empire from above; it grew from below." [1]

Shades of Hitler.

However, there were differences between Hitler-allegiance and emperor worship. Hitler's rise to power was rapid in comparison. Emperor worship was a gradual development which grew out of the gratitude of the provincials for what Rome had done for them. When Rome took over a country and unpredictable tyrants were booted out, Roman justice was established. The Roman peace, *Pax Romana,* was unlike anything the world had seen and people were deeply grateful.

It was not enough for the populace to appreciate Rome. This was an impersonal thing. The spirit of Rome needed to be personalized; the emperor of Rome began to be regarded as divine.

Before Christ was born, Caesar was worshiped.

"The first temple to be erected to the godhead of the emperor was built in Pergamum in 29 B.C. Caesar worship had begun." [2]

Can you imagine what happened in Rome? The empire was vast, it had many races and languages. It needed a unifying principle and "religion" can be a very unifying influence. Soon every Roman citizen was compelled to burn a pinch of incense and say, "Caesar is Lord."

Is it any wonder that Caesar worship ran head on into Christianity? When the Christians refused to call Caesar their lord, they were subjected to inhuman persecution. The gigantic movie epics showing the lions in the arena with the Christians and the public killings for the gory amusement of the Roman pagan leaders are not conjured from the imaginations of Hollywood producers.

There are many seeds which are planted in the breeding ground for dictators; anarchy, lawlessness, moral decadence, human desperation, and false hero worship fertilize the fields that produce despots. All the gold carat and two-bit demagogues of history grew out of the soil of the times.

Where Do We Fit In?

Are we living in a peaceful, placid era when people enjoy an environment free of tensions? Ridiculous question to ask, isn't it? But when we slow down our daily lives long enough to take a hard, realistic look at the generation we live in, it's a real shocker.

There are many shoulder-shruggers who say, "Always been crime and wars—always will be—why get all hot and bothered about what's going on today?"

A short time ago we saw a graph in a news magazine which indicated the climb in serious crimes in the United States from 1960 to 1968. If you had been an ant on that page you would have had a very steep stairway to climb each one of those eight years. While the number of crimes in America was increasing 122 per cent, the population rose only 11 per cent.[3]

Many people have stopped talking about the "crime rate." They now refer to the "crime epidemic."

Crime is waged on a grand scale by nations. Since World War II the world has been embroiled in conflict that seems to grow increasingly vicious. There has been a rebirth of guerrilla warfare; revolutions and revolutionary movements are becoming a way of life during this latter part of the twentieth century.

War and more war. Has there ever been a time when

the potential for self-destruction was as great as it is today?

Another Boom

Some people feel that the concern about the population explosion is exaggerated. They point out that the marvelous technological advances of science will overcome the burgeoning boom in human beings on this earth.

They may be right. However, many experts who have studied and evaluated the population growth have arrived at statistics and conclusions which are rather awesome, to say the least. A 1969 report from the United Nations national policy panel on world population claimed that the population crisis is the world's concern and that it is as important as peace itself. This report projects the world population will reach 7.5 billion by the year 2000. Since the number of human beings on earth in 1968 was 3.4 billion, we see that if the estimate is correct, there will be more than twice as many people in the world thirty years from now.

This same report said "high fertility and high rates of population" contribute to pollution, congestion, urban sprawl, and a host of psychological ailments in developed countries and might mean widespread famine, increased illiteracy, unemployment, squalor, and unrest threatening the foundations of public order in developing countries.[4]

Chairman of the Genetics Department at Ohio State University J. Bruce Griffing is quoted as saying that "Unless mankind acts immediately, there will be a worldwide famine in 1985, and the extinction of man within 75 years."[5]

We may not think this will affect us personally, but in 1967 Dr. Stanley F. Yolles, director of the National Institute of Mental Health, said:

"Population movement and the increased pressures of a speeded-up society undoubtedly are causing an increasing amount of emotional disturbance. Between 1960 and 1965 the number of U.S. psychiatrists, psychologists, and other mental health workers rose forty-four per cent. Youngsters are being admitted to mental hospitals in numbers seven times their share of the total population. Many of those left on the outside proclaim loudly their 'alienation from society.' "[6]

If we complain about not being able to find "breathing space," or resent being squeezed into an impersonal computerized society, imagine what it might be like thirty years from now, if we're still around!

Men who are studying population biology, such as Paul Ehrlich, professor at Stanford University and expert in this field, are inclined to be doomsters because of the research they have done. Ehrlich, for instance, says: "Mankind may be facing its final crisis. No action that we can take at this late date can prevent a great deal of future misery from starvation and environmental deterioration."[7]

All Systems Go

It doesn't take a "religious" person to discern the fact that what is happening is setting the world in the proper frame for a dictator. We see anarchy growing in every country. We see established standards of morality thrown aside for a hedonistic brand which is attractively labeled the "New Morality." We see the super-weapons and the threats of atheistic leaders in world powers who would not hesitate to use those weapons if they would further their drive for conquest.

A view seems to be creeping into the consciousness of concerned people that the problems and the tensions of the world need to be controlled by a "strong hand from someplace."

Even Arnold Toynbee, the eminent historian, said on a radio broadcast that "By forcing on mankind more and more lethal weapons, and at the same time making the world more and more interdependent economically, technology has brought mankind to such a degree of distress that we are ripe for the deifying of any new Caesar who might succeed in giving the world unity and peace."

Who Is the "Future Fuehrer"?

The time is ripe and getting riper for the Great Dictator, the one we call the "Future Fuehrer." This is the one who is predicted in the Scriptures very clearly and called the "Antichrist."

The Bible gives a perfect biographical sketch of this future world leader.

If you will follow this Scripture from Revelation, without being bothered by the figures of speech which are used, you will see that the Bible explains the meaning.

"And he stood on the sand of the seashore. And I saw a beast coming up out of the sea, having ten horns and seven heads, and on his horns were ten diadems, and on his heads were blasphemous names.

"And the beast which I saw was like a leopard, and his feet were like thosc of a bear, and his mouth like the mouth of a lion. And the dragon gave him his power and his throne and great authority" (Revelation 13:1, 2 NASB).

You may be saying that this doesn't sound like a human being to you, but just press on and follow carefully how the mystery unfolds.

Revelation continues with this description: "And I saw one of his heads as if it had been slain. . . ." Mark carefully the "as if" in the phrase above. ". . . and his fatal wound was healed. And the whole earth was amazed and followed after the beast; and they worshiped the dragon, because he gave his authority to the beast; and they worshiped the beast, saying, 'Who is like the beast, and who is able to wage war with him?' " (Revelation 13:3, 4 NASB).

This person, the Antichrist, is called the "beast" because from God's viewpoint that is exactly what he is. The passage is obviously talking about a person because the personal pronoun "he" is used. He is also described as a person of great authority.

In the seventeenth chapter of Revelation we are given the meaning of this description of the beast coming out of the sea. Revelation 17:15 says: "And he said to me, 'The waters which you saw where the harlot sits, are peoples and multitudes and nations and tongues' " (NASB).

The "harlot" refers to the religious system which will be tied in with this dictator. This will be described in more detail in the next chapter.

When it says that the beast will emerge out of the sea, it means that he will come out of the chaos of the nations.

In the Old Testament, Isaiah speaks of the chaos of the nations that is to come and says there is no peace for the wicked . . . they are like the "troubled sea," which is a symbolic picture of the Gentile nations (Isaiah 57:20, 21).

Like a Leopard, a Bear, and a Lion

"And the beast which I saw was like a leopard, and his feet were like those of a bear, and his mouth like the mouth of a lion" (Revelation 13:2 NASB).

To understand the meaning of this zoo which is described in Revelation, we go back to the predictive ministry of Daniel. As we studied in the previous chapter, Daniel describes his vision of four great beasts who come up out of the sea in succession. These beasts are the great Gentile empires which would rule the world. Daniel described these kingdoms using figurative expressions of wild animals.

Daniel 8 tells who the first, second, and third empires are; thus we can identify the animal figure by the order of the kingdoms. The first one of these empires which Daniel described was like a lion, which was the Babylonian Empire. He said the second great kingdom was like a bear, which was the Media-Persian Empire. The third kingdom would be like a leopard, and this was the Greek Empire.

Then Daniel said there would arise a fourth kingdom, which would take over what was left of the Greek Empire. That was Rome.

In Daniel 7:23, 24, Daniel predicts the fourth kingdom . . . and especially the person who would put it together. He says that the fourth kingdom (Rome) shall "devour the whole earth."

However, we saw in "Rome on the Revival Road" that there is a "phase 2" to the Roman Empire. Daniel says that out of the culture of the first Roman Empire ten kings shall arise, and another king after that who is different from the ten. This king is going to subdue three of the kings.

In other words, when this Roman dictator comes, he is going to take over the ten-nation confederacy. Seven of the kings or leaders will willingly give him their allegiance, but three of them will not. So he will overthrow these three leaders.

Now we are able to see how Scripture fits together. If you return to Revelation 13 and look again at verse one you will understand that the "ten horns" refer to this ten-nation confederacy and the "seven heads" are the seven leaders who form a coalition with the Antichrist.

Der Fuehrer's Personality Traits

Some people have said that they turn to the Book of Revelation to find out how it will all end . . . and then can't understand a thing that is said. Now that we know, however, who the leopard, the bear, and the lion were, we should begin to see what Revelation 13:2 means.

It is said here that the beast (the Roman dictator, the Future Fuehrer, or the Antichrist) will be like a leopard, a bear, and lion. A leopard is quick to seize his prey. The leopard was the Greek Empire. Alexander the Great, the Greek military genius, was known for the speed with which he overcame his enemies. He was fearless and strong, a human counterpart of the great leopard. Alexander was a world conqueror who led his armies to the outermost fringes of the known world of his time.

"A bear" referred to the Media-Persian Empire, which was conquered by Alexander. This kingdom, similar to the animal it exemplifies, was very strong and powerful.

The lion (Babylon) is regal. The way he walks and holds his proud head has class. Babylon was an elegant monarchy. Its palaces were splendid and the famous "hanging gardens of Babylon" were considered one of the seven wonders of the world.

Now we are beginning to see, little by little, the picture emerging of the Future Fuehrer. His conquest will be rapid, he will be very strong and powerful, and there will be an air about him which is self-assured and proud.

It is important to note as we follow in Revelation 13 that "the dragon gave him his power and his throne and great authority" (verse 2b).

Who is the dragon?

If we turn back to Revelation 12:9 we are told who he is.

It says, "And the great dragon was thrown down, the serpent of old who is called the Devil and Satan" (NASB).

We read that the dragon is going to give the Antichrist his power. In other words, Satan himself is going to give him fantastic power. He is going to be able to work all kinds of miracles. This is one reason that Christians should not get too excited when they see a miracle. It may not be a miracle of God. Satan is a miracle-worker and he has been able to work miracles from the beginning.

But there is going to be a time when Satan is allowed to work all kinds of supernatural acts through men. It is said that Satan is going to send this man, his masterpiece, with all sorts of signs and wonders and miracles (II Thessalonians 2:9).

The second thing that will be given this man is a throne. This means world government. This throne was offered to Jesus Christ. In Luke 4 it is told how Satan came to Christ during His temptation in the desert and offered Him all the authority and glory of the world. Satan said He could have all the kingdoms to rule, if He would fall down and worship him. There is only one man who could resist that kind of an offer. Satan offered Him a crown if He would by-pass the cross. But Jesus wouldn't buy it.

Now we are studying a man who will accept this throne wholeheartedly. He will be worshiped as Satan is worshiped, with forms of idolatry that we can only guess in our wildest leaps of imagination.

All around the world today the increase in the mystic, occult, and even devil-worship is so pronounced that people are beginning to question what it's all about. There are churches in some of the major cities of America which actually incorporate into their "religious ceremonies" the worship of the devil. A newspaper report said, "Moving hand-in-hand with astrology in some cities is a rising interest in witchcraft and hypnosis. Courses in the art of the history of witchcraft have popped up at a few recognized colleges and at most 'free universities' (where students design the curricula.)" [8]

A university student told this writer that he knew of several people who had signed up for a college course on witchcraft believing that it was a "how to" course, and after discovering that it was more history than mystery, dropped out of the class.

However, the Satan-worship which will be initiated at the time of the world reign of the Future Fuehrer will make today's antics of the cultists look like nursery school.

A Mortal Head Wound

We are told in Revelation 13:3 that this great world leader will have a fatal head wound which will be miraculously healed. Many people have not known just what to make of this statement. Some have thought that what this means is that one of the empires of the ancient Roman Empire would be miraculously revived and brought back to existence. That is one possible interpretation. However, I do not believe that is the right interpretation. Here is why. . . .

Look for a moment at Revelation 13:14. This is speaking of the False Prophet, who will be an associate of the Great Roman Dictator. The verse says: "And he deceives those who dwell on the earth because of the signs which it was given him to perform in the presence of the beast, telling those who dwell on the earth to make an image to the beast who had the wound of the sword and has come to life" (NASB).

Whoever this person is with this fatal wound will have a statue made of himself, and men are going to worship this idol. You do not make an idol of an empire. You make an idol of a person.

The way in which this dictator is going to step onto the stage of history will be dramatic. Overnight he will become the byword of the world. He is going to be distinguished as supernatural; this will be done by an act which will be a Satanic counterfeit of the resurrection. This writer does not believe it will be an actual resurrection, but it will be a situation in which this person has a mortal wound. Before he has actually lost life, however, he will be brought back from this critical wounded state. This is something which will cause tremendous amazement throughout the world.

We could draw a comparison to the tragic death of John F. Kennedy. Imagine what would have happened if the President of the United States, after being shot and declared dead, had come to life again! The impact of an event like that would shake the world.

It is not difficult to imagine what will happen when this coming world leader makes his miraculous recovery. This man, the Antichrist, will probably not be known as a great leader until the time of his revival from the fatal wound. After that the whole world will follow him.

He will have a magnetic personality, be personally attractive, and a powerful speaker. He will be able to mesmerize an audience with his oratory.

"Who is like the beast, and who is able to wage war with him?" These are the expressions the people who live at the time of the appearance of the Antichrist will be saying. They will accept anyone who offers peace, since this is the great cry of the world.

What does this indicate? We recall that the *Pax Romana,* the Roman peace, was the reason the provincials willingly turned to Rome and eventually initiated Caesar worship. Law and order—peace and security—freedom from war. The same needs, the same desires were expressed in ancient times that the Bible says will be prevalent before the Antichrist begins his rule. He will be swept in at a time when people are so tired of war, so anxious for peace at any price, that they willingly give their allegiance to the world dictator who will promise them peace.

There is another place in the Bible where the words of the people who are alive at the time of the coming of the Antichrist are quoted: "While they are saying, 'Peace and safety!' then destruction will come upon them suddenly like birth pangs upon a woman with child; and they shall not escape" (I Thessalonians 5:3).

This is a direct quote from the people who will live under the false security of this world dictator. It is a cry of relief—a sigh of gratitude. "Peace and safety"—at last!

The mood of the world is being developed toward the acceptance of this person. Remember what Toynbee said: "We are ripe for the deifying of any new Caesar who might succeed in giving the world unity and peace."

I Am God

The Antichrist will deify himself—just like the Caesars did. He will proclaim himself to be God. He will demand that he be worshiped and will establish himself in the Temple of God (II Thessalonians 2:4).

There is only one place where this temple of God can be and that is on Mount Moriah in Jerusalem, on the site where the Dome of the Rock and other Moslem shrines now stand. There are many places in the Bible that pinpoint this location as the one where the Jews will rebuild their Temple.

The Antichrist, who is called by many names in the Bible, but in II Thessalonians is called the "lawless one," will come in on a wave of anarchy. This is why the world will be ready to receive him.

You may be asking, "How can this possibly happen while Christians are in the world?"

It can't. As we understand II Thessalonians 2:6–12, the restraining power of the Spirit of God, within believing Christians, will hold back the rise of this World Dictator. We believe it is not until this "restrainer" is removed that his power and might will be exerted over the entire world.

We shall explain this in detail in the chapter on "The Ultimate Trip."

Authority to Act for Three and One-half Years

"And there was given to him a mouth speaking arrogant words and blasphemies; and authority to act for forty-two months was given to him" (Revelation 13:5 NASB).

Forty-two months is three and one-half years and this is the three and one-half years prior to Christ's personal, visible return to this earth. This period of time will make the regimes of Hitler, Mao, and Stalin look like Girl Scouts weaving a daisy chain by comparison. The Antichrist is going to be given absolute authority to act with the power of Satan.

"And he opened his mouth in blasphemies against God, to blaspheme His name and His tabernacle, that is, those who dwell in heaven" (Revelation 13:6 NASB).

This is interesting. Why would he "blaspheme" or "bad-mouth" those who will dwell in heaven? And who are the ones who dwell in heaven; why would he even bother with them? You and I are the ones who are going to dwell in heaven, if we are true believers in Jesus Christ. If we are gone from the earth at this period, which this writer believes the Scriptures prove, how is this fellow going to explain the fact that a few million "religious kooks" are missing? He will, of course, have to discredit the Christian belief, explain it away, or "blaspheme" it and the believers.

"And it was given to him to make war with the saints and to overcome them . . ." (Revelation 13:7a NASB).

It is logical to ask at this point, how is he going to make war with the saints when they are gone from the earth? "The saints" are the people who are going to believe in Christ during this great period of conflict. After the Christians are gone God is going to reveal Himself in a special way to 144,000 physical, literal Jews who are going to believe with a vengeance that Jesus is the Messiah. They are going to be 144,000 Jewish Billy Grahams turned loose on this earth—the earth will never know a period of evangelism like this period. These Jewish people are going to make up for lost time.

They are going to have the greatest number of converts in all history. Revelation 7:9–14 says they bring so many to Christ that they can't be numbered.

However, the Antichrist is going to unleash a total persecution of these people.

"And it was given to him . . . authority over every tribe and people and tongue and nation . . ." (Revelation 13:7b NASB).

He will be the absolute dictator of the whole world! This is the Future Fuehrer.

Who Will Worship Him?

"And all who dwell on the earth will worship him, every one whose name has not been written from the foundation of the world in the book of life of the Lamb who has been slain" (Revelation 13:8 NASB).

The book of life to which this passage refers is for those who put their personal faith in Jesus Christ. Christ is the "Lamb" who was slain.

The Old Testament predicted the coming of Jesus in hundreds of prophecies. Over 300 specific predictions were fulfilled in His life and death. He is conclusively proven as the only One who could be the Messiah.

He has a book. That book contains the names of everyone who simply puts his faith in Jesus for having paid the penalty for his sins. Throughout the history of the Jewish people the Passover Lamb typified a sacrifice. What an incredible thing that so many Jews would not see Jesus as the personification of the Lamb. He was the real Lamb of God.

When John the Baptist, one of the greatest prophets of all time, first saw Jesus coming toward him, he said, "Behold, the Lamb of God that takes away the sin of the world" (John 1:29 NASB). In this statement he summarized the whole significance of the Old Testament. It all pointed to Jesus.

So we see who will worship the Antichrist. Everyone will worship him who has not put his faith in Christ.

Is it possible for a moment to project our thoughts toward that time when the entire world will look upon one human being as the supreme leader? The Antichrist will need a lot of help to carry out his diabolical schemes. His staunchest ally will be

The False Prophet

In Revelation 13:11–18 we are introduced to this infamous character. This person, who is called the second beast, is going to be a Jew. Many believe he will be from the tribe of Dan, which is one of the tribes of the original progenitors of the nation of Israel.

The False Prophet (he is called that in Revelation 19:20 and 20:10) will be a master of satanic magic. This future False Prophet is going to be a devilish John the Baptist. He will aid and glorify this Roman Dictator; he will proclaim him the savior of the world and make people worship him as God.

It is logical to ask how the False Prophet will force this worship of the Roman Dictator. He will be given control over the economics of the world system and cause everyone who will not swear allegiance to the Dictator to be put to death or to be in a situation where they cannot buy or sell or hold a job. Everyone will be given a tattoo or mark on either his forehead or forehand, only if he swears allegiance to the Dictator as being God.

Symbolically, this mark will be 666. Six is said to be the number of man in Scripture and a triad or three is the number for God. Consequently, when you triple "six" it is the symbol of man making himself God.

The cleverness of this economic vise is ghastly to contemplate. But no one could ever accuse Satan of not being clever.

O Come Now . . .

Do you believe it will be possible for people to be controlled economically? In our computerized society, where we are all "numbered" from birth to death, it seems completely plausible that some day in the near future the numbers racket will consolidate and we will have just one number for all our business, money, and credit transactions. Leading members of the business community are now planning that all money matters will be handled electronically.

Upstage

We believe that the dramatic elements which are occurring in the world today are setting the stage for this magnetic, diabolical Future Fuehrer to make his entrance. However, we must not indulge in speculation about whether any of the current world figures is the Antichrist. He will not be known until his sudden miraculous recovery from a fatal wound.

There would be no earthly advantage in being alive when the Antichrist rules. We believe that Christians will not be around to watch the debacle brought about by the cruelest dictator of all time.

Beloved, do not believe every spirit, but test the spirits to see whether they are from God; because many false prophets have gone out into the world.

I JOHN 4:1 (NASB)

10

REVIVAL OF MYSTERY BABYLON

It is not our purpose to be shocking or offensive. The prophecies of the Bible are a vital part of God's Word, but should not be used for sensationalism. Some of the symbolism in this chapter may seem very strange to modern ears, and some of the indictments of existing beliefs may trigger the defense mechanism in some readers. However, when God's Word is clear, it was never intended that man should dilute what He has to say.

All through the Scriptures we find that Christ dealt strongly with the religious leaders and false prophets who put on their many-colored coats of righteousness and led people astray. Jesus called them hypocrites, fools, and vipers, not exactly soft terms. He tipped over the tables in their places of so-called worship, which was not what men today might consider being broad-minded.

We are told in the Bible that before the seven-year period of tribulation there will be an all-powerful religious system which will aid the Antichrist in subjecting the world to his absolute authority. For a time this religious system will actually have control of the Dictator.

There are several names given to this one-world religion, all of them drawing a perfect analogy in meaning. It is called the Great Harlot, or prostitute; the "harlot" represents a religion which prostitutes the true meaning of being wedded to Christ, and sells out to all the false religions of man.

"Babylon" is another term used for the one-world

religion. It is easy to see how a place could also be a system when we consider contemporary examples: Broadway means the theater today, Madison Avenue draws the mental image of the slick advertising world, Wall Street is the hub of finance.

'Tis a Puzzlement

As pieces of the prophecy puzzle appeared to fall into place there was one important part that was lost to me. The Bible outlines very specifically that there would be a one-world religion which would dominate the world in the time before the return of Christ. However, this seemed so remote, with so many different religions competing for the minds and hearts of men, how on earth could people unite in allegiance to just one religion?

Five years ago, for instance, as I surveyed the college campuses where I ministered, it seemed that most of the intellectual community was alienated by any concept of the supernatural. Many of them were not receptive to any form of "religion," which was considered a crutch for the weak, or a rigid set of rules not worth trying to follow.

We should be careful to note at this point that we are using the term "religion," not Christianity. Christianity is not a religion. Religion is the process of man trying to achieve goodness, perfection, and acceptance with God by his own efforts. Christianity, on the other hand, is God taking the initiative and reaching for man. Christianity is God saying that man cannot reach Him except through the one path He has provided—through the acceptance of His Son, Jesus Christ.

But the scene has changed rapidly in the intellectual community in the past few years. Many of those who scoffed at "religion" have become addicted to the fast-moving upsurge in astrology, spiritualism, and even drugs. What does all this mean? Does it have any significance in Biblical prophecy?

We believe that the joining of churches in the present ecumenical movement, combined with this amazing rejuvenation of star-worship, mind-expansion, and witchcraft, is preparing the world in every way for the establishment of a great religious system, one which will influence the Antichrist.

To understand the significance of these movements in relation to this one-world religious system, we should study carefully what the Bible has to say about them. As we begin to examine this in depth, there emerges a pattern which makes the Bible more up-to-date and relevant than any book today.

Origin of Astrology

Many popular American magazines have been commenting recently on astrology and its impact on the contemporary scene; most of these secular writers have done a good job of historical research into the subject. However, few touch upon the best history book of all in doing their reference work.

In Genesis we find that astrology got its beginnings in Babylon, which is verified by secular writers in their historical accounts. There is a remarkable statement in Genesis 11 about the first astrological observatory.

After the great Flood, the Bible tells us, the whole earth had one language. The people who were alive at that time settled in the land of Shinar, which is near the meeting of the Tigris and Euphrates rivers. This was ancient Babylonia, one of the earliest centers of civilization.

"And they said, Go to, let us build us a city and a tower, whose top may reach unto heaven . . ." (Genesis 11:4 KJV).

Now the city they were speaking about was Babylon, and the tower was the famous Tower of Babel. The interesting thing is that the word "tower" is a word that can mean ziggurat, which was the ancient observatory from which the priests would gaze at the stars.

In many versions of the Bible you will find that in this verse from Genesis 11 the words "may reach" are in italics. This is because they were not in the original Hebrew writings of the Bible. When this passage is studied in Hebrew it becomes obvious that these people had enough intelligence and know-how to realize that they could not build a tower which would actually reach to heaven. It's true that the Babylonian builders were geniuses, but they were not stupid. A tower with its top in heaven meant it would be used to study the stars, chart their courses, and make predictions. Henry H. Halley, who has assembled the comprehensive *Halley's Bible Handbook,* said that the whole purpose of the ziggurats was idolatrous worship.[1]

This origin of astrology was described in what may have been among the first writings known to man, a form of hieroglyphics devised by the ancient race of the Chaldeans who began to give the stars certain meanings. They divided the heavens into the twelve sections of the Zodiac and said that the stars control the destiny of men. This religion started to flourish and have its greatest glory of history in the Babylonian Empire, which was primarily made up of the priestly caste of the Chaldean people. These Chaldeans became thc aristocracy of the priesthood. We know how esteemed these astrologers and magicians were because every king in Babylon built giant ziggurats for them. The astrologers were considered almost equal in power to the king.

We see God's attitude toward this star-worship when we are told that He destroyed this Babylonian observatory. Nimrod, the first world dictator, was the leader in the Tower of Babel enterprise and through this religious system established at Babylon undoubtedly wished to pursue the ambition of all dictators, and that is to have a one-world government. He did not take God into account, however. God says when you put everyone under one dictator there would be no evil which could be restrained. When the Tower of Babel was destroyed, it was a turning point in history. Where there had been but one language, now God initiated many languages and scattered the people throughout the earth.

This passage shows that God's plan for the world until the Prince of Peace returns is not an international one-world government, but nationalism. This is the one way the world can keep from falling under a dictator who could virtually destroy mankind.

The Scripture says that a Great Dictator is coming and he will be boosted to power, and strengthened in his grasp upon the world with the assistance of the ancient religion called Mystery, Babylon. This is the very religion which started in the Genesis account and made possible the first world dictator.

Some clairvoyants today, without knowledge of Bible prophecy, are saying the same things that the Scriptures tell us. In a publication which specializes in psychic phenomena there was an astounding article which said that "History and many signs of the times point to man's preparation for the coming new world leader." The writer in this magazine dealing with extrasensory perception, prophecy (not Bible), and spiritual healing, described various events which would precede this new "Leader," among which were . . . "Conditions shall change and, in the due course of time, people shall be led into a new Age where they have different abilities from those now existing. There shall be clairvoyance and telepathy as there was before the so-called, miscalled, Tower of Babel, in which through the abuse of special powers mankind lost its telepathic abilities *for the time being.*"[2] (Emphasis is ours.)

Mystery, Babylon, Condemned

The great prophet Isaiah, in chapter 47 of his book, gives an analysis of the Babylonian worship and religion. The description and the judgment to be brought upon this "religion" is so significant for today that we can scarcely contain our continuous excitement over prophetic pictures in the Bible.

Isaiah, speaking predictively in 47:1, says: "Come down, and sit in the dust, O virgin daughter of Babylon, sit on the ground: there is no throne, O daughter of the Chaldeans: for thou shalt no more be called tender and delicate" (KJV).

A glance down to verse 5 will indicate further condemnation of Babylon: "Sit thou silent, and get thee into darkness, O daughter of the Chaldeans: for thou shalt no more be called, The lady of kingdoms" (KJV).

(Literally, the "lady" means the "Queen of kingdoms." When we study Revelation 17 this is going to be a very important figure of speech.)

The prophet goes on to analyze the wickedness of

Babylon and the reason why God is going to judge this empire. Isaiah shows how Babylon grew away from the true God and was involved in worldly pleasures. Wisdom and knowledge perverted Babylon, and the prophet says that desolation shall "come upon thee suddenly."

However, the most important thing to see is that Babylon was entangled in "sorceries" and "enchantments," two important words which we shall examine carefully . . . and that this civilization could not be saved by "the astrologers, the stargazers, and the monthly prognosticators" who were the advisers to the Babylonians.

Stay Tuned to the Facts

We are not playing a game of Biblical hopscotch when we turn from one prophet to another. If there was not a strong recurring theme in all the prophets, this book would have no validity. It is the complete agreement of all parts of Biblical prophecy which makes its study so absorbing, so filled with tingling vitality for the twentieth century that it is a wonder that every preacher in every pulpit doesn't shake the ecclesiastical rafters with this subject.

Daniel was a prophet whose themes are far-reaching and who predicted events which are yet to happen in the history of our world, events which we are studying in the context of this book. Daniel was a Hebrew nobleman who was taken captive in Babylon by the king of the empire, Nebuchadnezzar. This king had Daniel and several other Jewish noblemen who showed unusual intelligence put into a special school of the wise men. The "wise men" was a general designation given to the astrologers and magicians of the Babylonian empire and Daniel was trained in all the wisdom of the Babylonians. He became an expert in these fields, but he didn't buy it. It was a tremendous test of Daniel's faith, because he was literally brainwashed by the Babylonian religion, but he kept his belief in God, in spite of all this training.

Nebuchadnezzar had a dream which really bothered him and gave him insomnia. He called the whole gang out to help him with this—all his "magicians and astrologers and sorcerers and Chaldeans" (Daniel 2:2, 10, 27).

The first category or word which is translated "magician" means the special sacred scribes—an order of the wise men who had charge of the sacred writings which were handed down all the way from the time of the Tower of Babel. Some of the earliest-known literature on the face of this earth were these books of magic, astrology, and the black arts.

The next word, "astrologer," means literally to whisper—it came to mean "conjure." These were men who were enchanters who would cast spells through the spirit medium. They had all kinds of chants and low, muttering songs through which they thought they could cast out evil.

The "sorcerers" means the black magic art. The same word is used for the Egyptian magicians who confronted Moses (Exodus 7:11). Through their black art they duplicated several miracles which Moses performed in the court. For instance, they turned water into blood. When Moses cast down his rod and it turned into a serpent these "sorcerers" did the same thing in the court of Pharaoh. It wasn't until God enabled Moses to perform miracles that they couldn't duplicate that these magicians said that this was the finger of God. The Egyptians had all of these arts which the Babylonians exported to them.

The next word, "Chaldeans," was the priest caste of all of them. Wherever you see the word "Chaldean" it can be translated equally well as "astrology." Several great linguists unanimously agree on this point that the Chaldeans studied a person's birthday, asking the very hour a person was born, and then they would cast a horoscope of his destiny. You will find that this thinking has permeated the Aramaic world—the Arabs, the Persians, etc. This is where the idea of Kismet came from; it means it is inevitable, it is fate—what will happen will happen. The ancient astrologers believed that your fate was written in the stars before you were born and you can't alter the course of your destiny.

The Babylonian kings did not make a move without consulting their astrologers, and the Media-Persians who conquered the Babylonians did the same thing. They, too, had their court astrologers and wise men. When the

Greek, Alexander the Great, conquered the Media-Persian Empire he also began to follow the astrologers and their wisdom. The practice was transported to Rome where we see the Caesars consulting the augurs, who were experts in astrology, spiritism, and the black arts.

When Nebuchadnezzar called upon his wise men for help and advice, none of them could tell him his dream or the interpretation. Daniel, however, proved he was a true prophet of God when he was able to tell the king what the dream was and then give its interpretation. Daniel really puts the bite on his fellow colleagues and shows them up for the frauds they were.

Nebuchadnezzar's dream, described by Daniel and interpreted by this prophet of God, turns out to be the whole course of world powers that would conquer the world right up to the second coming of Jesus Christ.

Daniel shows that astrologers cannot accurately predict the future. As a matter of fact, God condemns astrology. In Deuteronomy 18 astrologers were commanded in that day under the law of Moses to be put to death. Astrology is the backbone of the ancient religion of Babylon, but there is no such thing as a man's fate being controlled by the stars.

Man's fate is controlled by God alone.

Whenever a person gets sold out by things like astrology, it will be like the Israelis. Israel was judged and virtually destroyed and taken to Babylon in 606 B.C. because the people became devoted to idolatry, especially astrology. II Kings 23 makes this clear when it says that God wiped out the high places and all of the places where they burned incense to the stars of heaven, or to the constellation (the twelve signs of the Zodiac).

Emergence of the One-world Religion

Now we are coming to some of the most important pieces of the prophetic puzzle which are shown in the Scriptures. In Revelation 17 the apostle John has a vision which shows the future and precisely what is going to happen on earth the last seven years before

Christ returns. In Revelation 17 John is given one of the most important prophecies for us to understand because he is exposing a one-world religious system which will bring all false religions together in one unit. Through this system Satan's Antichrist will take over the world —and he is going to do it first with Rome as home base and then from Jerusalem.

Scarlet O'Harlot

Look what John says in Revelation 17:3–5: "And he carried me away in the Spirit into a wilderness; and I saw a woman sitting on a scarlet beast, full of blasphemous names, having seven heads and ten horns. And the woman was clothed in purple and scarlet, and adorned with gold and precious stones and pearls, having in her hand a gold cup full of abominations and of the unclean things of her immorality, and upon her forehead a name was written, a mystery, 'BABYLON THE GREAT, THE MOTHER OF HARLOTS AND OF THE ABOMINATIONS OF THE EARTH' " (NASB).

It is sometimes difficult for the Bible reader to grasp this symbolism. Understanding "the mother of harlots," or the great harlot, is important to the basic understanding of the Christian belief, versus all religions. This church which the prophet describes claims to be united to Jesus Christ in mystical marriage, but is really an adulteress in the spiritual sense. This church professes an allegiance to God, but worships a false religious system.

The main thrust of this passage in Revelation 17 is about Mystery, Babylon, who is typified by this woman. "And I saw the woman drunk with the blood of the saints, and with the blood of the witnesses of Jesus. And when I saw her, I wondered greatly. And the angel said to me, 'Why do you wonder? I shall tell you the mystery of the woman and of the beast that carries her, which has the seven heads and the ten horns' " (Revelation 17:6, 7 NASB).

We know the meaning of the "beast," for we studied

this in the chapter on "The Future Fuehrer." The ten horns, as we have seen, refers to the revived Roman Empire, or a ten-nation confederacy.

The fact that this woman, who is the Mystery, Babylon, is riding upon this beast shows that she controls him. Historically this "religion" has controlled many empires. Notice what John the Apostle, writing in about A.D. 95, says about these empires which the "woman" has controlled: "Here is the mind which has wisdom. The seven heads are seven mountains on which the woman sits, and they are seven kings; five have fallen, one is, the other has not yet come; and when he comes, he must remain a little while" (Revelation 17:9, 10 NASB).

We must look at this from the apostle John's perspective. He speaks of seven kingdoms, five of which have fallen. We must examine what five kingdoms over which this mystery religion of Babylon exerted authority have fallen. The first great kingdom where the Babylonian religion had great sway was the Chaldean. The second world power which was virtually controlled by this religion was Egypt. The great Egyptian pyramids had astrological significance, as well as being a burial place for the kings, and the sphinx is the key to where the earth starts in the twelve sections of the Zodiac. The head of the sphinx is a woman, with the body of a lion. The word "sphinx" means to "join together" in Greek—it shows that the first part of the circle of the zodiac is the woman, Virgo, and the last part is the lion, Leo. What the sphinx does is to put the two together and show the beginning and the end of the Zodiac.

The next great world kingdom to be controlled by the mystery religion was, of course, Babylon itself under the great Babylonian rulers. The fourth kingdom was the Media-Persian Empire, and the fifth was the Greek Empire.

John says that "five have fallen, one is." The kingdom which was present at the time John wrote this was Rome. The mystery religion of Babylon was in Rome, exerting its great influence over the decisions of the empire. The other kingdom which "has not yet come" is referring to the revived Roman Empire.

The Antichrist will come up out of the culture of the ancient Roman Empire. He will come to the ten-nation confederacy, take it over, and make it an eighth form of power.

We believe that we are seeing, with all of the other signs, the revival of Mystery, Babylon—not just in astrology, but also in spiritism, a return to the supernatural, and in drugs.

More insight into this last one-world religion is given in Revelation 9:20: "And the rest of mankind, who were not killed by these plagues, did not repent of the works of their hands, so as not to worship demons, and the idols of gold and of silver and of brass and of stone and of wood, which can neither see nor hear nor walk" (NASB).

Some may scoff at the idea that idol worship will become prevalent. Even in America, however, there are growing cults which actually do worship stone and metal idols. At one of California's colleges some young men were seen at dawn, indulging in a primitive sort of sun-god worship on the lawn of a fraternity house. When they were questioned about their rather strange activities they said, "There's a bunch of us up at Big Sur who worship idols and stuff."

And what else will be seen during these times preceding Christ's return? "And they did not repent of their murders nor of their sorceries nor of their immorality nor of their thefts" (Revelation 9:21 NASB).

One of the words here is extremely important—the word "sorceries." It comes from the Greek word *pharmakeia,* which is the word from which we get our English word pharmacy. It means a kind of occult worship or black magic, associated with the use of drugs. This word is mentioned several times in the Book of Revelation. It is said of the great religious system that "all the nations were deceived by your sorcery" (Revelation 18:23).

If there is anyone today who is not conscious of the spread of drug addiction, particularly among college and high school students, and now reaching down into the junior high and even the grade school level, then they are blind to what is happening. The increase in the use of narcotics and all forms of dangerous drugs has spiralled to such an extent that statistics written today would be out of date by the time this book is published.

A newspaper we saw recently quoted narcotics officials in the juvenile division of a police department saying that they were so alarmed that they thought the situation should be labeled a "Drug Epidemic" and treated by drastic and swift measures before it was more out of hand than it is now. One officer said that whatever they say about the arrests of young people on drug charges is like an iceberg. All they are able to see is what protrudes above the surface. Lurking beneath the known dangers are thousands and thousands of users and pushers that haven't been caught.

I used to wonder how on earth people could see the supernatural things which would occur during the Great Tribulation, the seven-year period before the return of Christ, and still not turn to God. I have seen a graphic example of how minds can be clouded from a person I met in a fraternity house at one of our major universities. He was an outstanding young man and expressed the desire to know more about Christ. I met with him for several weeks and he said, "I believe it, but I just don't want to commit my life to Christ."

When I met him a few months later he looked completely different. He said, "I'm really religious now—I feel sorry for you—I've been taking trips and I've really seen God. Only this God is the King of Darkness—this is the one we worship."

This man had completely blown his mind. He is at this point, apart from a miracle, completely beyond reach. Satan uses these hallucinatory drugs to take man to a deeper level of approach with him. You talk to some people who have been on drugs for a long time and they will tell you, "I know the devil is real—I've seen him."

We believe these drugs reduce a man's thinking and mentality to a point where he is easily demon-possessed. Demons are under the control of Satan and the Bible speaks about them in abundant terms, so we're not talking about spooky things.

There are all sorts of new spirit groups which are worshiping Satan. A newspaper feature article told of "Modern Witches: Old Black Magic but a New Spirit." The story concerned the upsurge in witchcraft in England and told how people take it very seriously. "So ostensibly do the hundreds, perhaps even a few thousand, of witches in Britain who are riding again since the centuries-old Witchcraft Act was repealed in 1951. Covens, assemblies of witches are even said to be thriving in Los Angeles and New York."

As one modern witch describes, "For us it's our religion—we worship a horned god, the prince of darkness, and this makes some people say we're devil-worshipers." [3]

A major television station had a potpourri of news and showed the great interest high school students have in witches. The commentator said, "Nearly every respectable high school these days has its own witch." The program pictured a pretty 16-year-old, eyes staring and voice chanting incantations, conjuring up all sorts of spells. A psychiatrist intoned that he thought the trend toward witchcraft was "healthy" for some because it helps relieve aggressive feelings.

The upsurge of interest in astrology was documented in the first chapter of this book. Man seems to be on this great quest for knowledge and assurance of the future with such intensity that all types of mystic ideas are being accepted today.

Where Is the Church?

We have developed the various forms of religious expressions which we believe constitute a revival of Mystery, Babylon, but have not included the organization which the Bible says will be a definite part of this one-world religion. This is the visible church which is characterized by increasing unbelief and apostasy.

This description of the visible or physical church is given with sadness. We find it difficult not to be disheartened when we see what is happening in the churches which call themselves "Christian." However, if we follow Bible prophecy carefully we see that it is made absolutely clear that in the time preceding the return of Christ there would be a "falling away" from the basic doctrines of Christ by the churches. Some churches are not just falling away, they are plunging recklessly toward destruction with their disbelief and blasphemous programs.

It's not uncommon to hear someone say, usually in a state of anger over some real or imagined injustice, "Why doesn't the church do something about it?"

It is imperative for us to have a clear understanding of who and what "the church" is before we can understand its apostasy. ("Apostasy" means an abandonment or desertion of principles or faith.)

The apostate church is, always has been, and will be, the visible, physical gathering of people who may call themselves Christians. These churches may be of any denomination. They may be magnificent cathedrals or store-front missions, a congregation of thousands or a gathering of a few. However, no matter what sacred or holy name is applied to the visible church, this is no guarantee that it teaches and preaches the truth of God.

The true church, on the other hand, includes all believers in Christ. Many places in the New Testament speak of those who are joined to Christ, the Head of the body. In other words, Christ is the Head of the true church.

"He is also head of the body, the church" (Colossians 1:18 NASB).

"For even as the body is one and yet has many members, and all the members of the body, though they are many, are one body, so also is Christ. For by one Spirit we were all baptized into one body, whether Jews or Greeks, whether slaves or free, and we were all made to drink of one Spirit" (I Corinthians 12:12, 13 NASB).

An organized church may have among its members those who are true believers and those who merely claim to be believers in Christ. We must not make the mistake of saying that everyone who belongs to a church which has strayed far away from the teachings of the Bible is a non-believer. A true believer can be surrounded by apostasy in the church or denomination or the council of churches to which he belongs. Usually, if the believer is aware of what is happening at the policy-making level of his church, he is pretty miserable—or angry—or disgusted.

Some of you may be thinking that every generation has seen this apostasy in the church. This is true, but the Bible says that as the countdown before Christ's return comes closer, the teachings of the false leaders of the church will depart farther and farther from God's Word.

"But false prophets also arose among the people, just as there will also be false teachers among you, who will secretly introduce destructive heresies, even denying the Master who bought them, bringing swift destruction upon themselves" (II Peter 2:1 NASB).

The Harlot

This one-world religious system is not described in delicate terms. A harlot, or a prostitute, is one who is unfaithful. She has prostituted her God-given womanhood. In the same manner, the church which says it belongs to God, but joins in worshiping through a false religious system, is prostituting its purpose.

How can we recognize the apostasy in the church today? What are the characteristics of this harlot?

Peter writes that in the "last days" there would be "mockers" who would say: "Where is the promise of His coming?" (II Peter 3:4 NASB).

John, the apostle of love, spoke strongly about false teachers who deny the fact that Christ will return bodily to earth the second time. He wrote: "For many deceivers have gone out into the world, those who do not acknowledge Jesus Christ as coming in the flesh. This is the deceiver and the antichrist" (II John 7 NASB).

We need to be alert. When we hear church leaders, teachers, or preachers questioning the visible return of Christ, this is a doctrine of apostasy.

The Bible teaches from beginning to end that man is basically sinful. Since this is contrary to so much of the humanistic teachings that surround us today, it is hard for some to swallow. But it does not make it any less true. "If we say that we have no sin, we are deceiving ourselves, and the truth is not in us" (I John 1:8 NASB).

There are many who admire Jesus of Nazareth as a great man, an outstanding teacher, but scoff at His deity. This is another form of apostasy. This is a denial of God, since the doctrine of the trinity is a basic tenet of Christianity. John says: "Whoever denies the Son does not have the Father . . ." (I John 2:23 NASB).

We have often had people say to us, "Sure, I believe in God, but you certainly don't think Jesus was God, do you?" Of course we do. The Bible also says Jesus was born of a virgin; to deny this is a denial of the miracles of God, what traditional Christianity has said throughout the centuries.

What is happening today in many of our standard brand denominations? One of the first exposés of the beliefs of our future ministers was made by *Redbook* magazine in August of 1961. The publishers hired one of the top pollsters in the nation to survey a full representation of our seminaries which are supposedly preparing men for Christian service in the Protestant churches. Here are some of the results—compare them carefully with what the Bible says about apostasy.

Of the ministers in training, 56 percent rejected the virgin birth of Jesus Christ, 71 percent rejected that there was life after death. 54 percent rejected the bodily resurrection of Jesus Christ. 98 percent rejected that there would be a personal return of Jesus Christ to this earth.

"But false prophets also arose among the people, just as there will also be false teachers among you, who will secretly introduce destructive heresies . . ." (II Peter 2:1 NASB).

We are seeing "destructive heresies" bought in wholesale lots. In the 1968 annual of the *Encyclopaedia Britannica* in a special section on religion the writer wrote that this marks a turn in the history of American theology, for not only are they saying that God is dead, but men who still claim to be Christian theologians are saying that there is no such thing as a personal God at all.

In the February, 1968, issue of *McCall's* a survey of major denominations showed that a considerable number rejected altogether the idea of a personal God. How, may we ask, can a man be a minister today if he does not believe in a personal God?

Although this trend has been accelerated in recent years, this has been going on for a long time. When people move away from Christianity the church will lose its power and influence to a great religious movement, a satanic ecumenical campaign.

The Ecumenical Mania

Years ago when we first heard about the ecumenical movement, we couldn't pronounce it, but we thought it sounded like a great idea. It seemed plausible that all the "good guys" in the churches should join together to fight all the evil on the outside. There are many fallacies in that way of thinking. When all of the various churches begin to amalgamate in one unwieldy body, soon the doctrinal truths of the true church are watered down, altered, or discarded. In their place we see political pronouncements and ecclesiastical shenanigans that astound the believer and repel the nonbeliever. The mass movement of the National Council of Churches and the World Council of Churches toward an umbrella-like structure which would cover all sorts of beliefs and camouflage its motives with "broadmindedness," seems to say: "Come unto me all ye that are weary and heavy-laden and I will give you—controversy!"

In May of 1969 the World Council of Churches, for instance, recommended that the churches should support violence if it is the last way to overthrow political and economic tyranny. This group also recommended that churches confess that they are "filled with blatant and insidious institutional racism." [4]

The news media carry stories every day about

churches being invaded by pressure groups who "demand" money or recognition. Churches are joining forces with those who oppose everything that is known as traditional Christianity. "Marxist-Christian dialogue" is very popular; not for the purpose of extending the love of God to the atheistic Marxists or Communists, but with the idea of exchanging "truths" and reaching a common ground of understanding.

When Gus Hall, one of the best-known spokesmen for the Communist Party in the United States, says that the current red goals for America are "almost identical" to those espoused by the liberal church, then perhaps it is time (really far past time) that people begin to wake up to what is going on in the apostate church.

"A big and growing trend toward unity is developing in many of America's churches at this time. That trend, some clergymen are saying, could lead to a 'superchurch' of immense religious and political power." [5]

Satan's real intention is not to have a godless political situation, he wants a religious situation. Satan loves religion, which is the reason he invades certain churches on Sunday. Religion is a great blinder of the minds of men.

Where Does Apostasy Lead?

Christ calls His followers, the true Christians, the salt of the earth. If the salt is removed this important preservative will no longer be able to stop the decaying process. When false teachings and doctrines become predominant there will be a decline in the moral climate, for the preservative has been taken away.

Hear what the Bible has to say about the time before Christ's return: ". . . in the last days difficult times will come. For men will be lovers of self, lovers of money, boastful, arrogant, revilers, disobedient to parents, ungrateful, unholy, unloving, irreconcilable, malicious gossips, without self-control, brutal, haters of good, treacherous, reckless, conceited, lovers of pleasure rather than lovers of God . . ." (II Timothy 3:1–4 NASB).

That is a strong indictment of the era in which we live, isn't it? Continuing in that passage, Paul says that learning or intellectualism will increase, but men will not come to the "knowledge of truth."

We have increased in technology so rapidly in the past few years that our grandfathers' heads would be spinning at what we take for granted. However, all the educational advances have not brought mankind one step further toward solving the basic needs of love, security, and true happiness; on the contrary, civilization seems more removed from these concepts than ever.

In this letter to Timothy, Paul also cites that there will be a pretense of worship, at the same time denying godliness. When we read of some of the magazines that fall under the category of "religious publications," we can understand what is meant by this pretense. God is not a living, vital reality, personified by Christ, He is hardly more than a "Thing," a "Reality," a "Ground of Being," a "Voice from Somewhere." When there is little to separate the church from a nightclub, a school, a social gathering, a political meeting, even a philanthropic group, then the salt has been taken away, its flavor is no different from any secular organization.

The Power Without the Salt

Although this great religious system of the time of the Tribulation, or the final seven-year period before the return of Christ, will be godless, it will also be powerful.

John says in Revelation 17 that this system would dominate the beast, or the Great Dictator who will be the head of the ten-nation confederacy. As we have said, it would seem that fear of Communism and the need for a common defense against the "King of the North" would drive the one-world political system into the arms of the "harlot" or the one-world religious system. The "harlot" will be "clothed in purple and scarlet, and adorned with gold and precious stones and pearls . . ." (Revelation 17:4 NASB). In other words, this religious system will be splendid on the outside, but corrupt to the core.

This harlot, as described in Revelation, is not only a system, but also a city. There is no question about where the city would be—it was Rome. It says that the woman sits (or rules) on seven mountains. Most elementary students of history or geography know that Rome is the city of seven hills. It is there that the religious system will reign for a time in coalition with the political system.

However, the Bible tells us about the end of this system. The political system and its dictator will hate this religious system after a time, because it controls the Antichrist and he wants to proclaim himself to be God, without any interference. The Great Dictator is not going to be a puppet; if strings are to be pulled, he wants to do the manipulating.

Revelation 18:2 shows that the destruction of the religious system will take place in two phases. "Fallen, fallen is Babylon the great." The first "fallen" refers to the destruction of the religious system by the dictator. This will take place in the middle of the Tribulation, at the end of three and one-half years. The second "fallen" refers to the sudden destruction of the city of Rome.

God-given Compassion

There is a tightrope which is walked by everyone who condemns the actions of churches, their leaders, or any belief which is no more than unbelief. One is accused of being "anti-church," "narrow-minded," or "dogmatic." However, in proclaiming the truths of God, revealed by the Bible and its prophets, we cannot dilute what is said. On the other hand, while loathing the actions and motives of any system which is religious, without knowing Christ, we do at the same time want to introduce persons within these systems to the only way to God, to Jesus Christ Himself. This is a distinction which is difficult for a non-Christian to understand, and sometimes equally difficult for the Christian to follow. It is against human nature to separate the man from the beliefs he appears to follow. We cannot do it alone. Only Christ, working in and through our whole being, can possibly give us compassion and love for what we ourselves have labeled unloving and despicable.

One small step for a man—one giant leap for mankind.
APOLLO 11 COMMANDER NEIL ARMSTRONG
20 JULY 1969

11
THE ULTIMATE TRIP

And the world caught its breath. Science fiction had prepared man for the incredible feats of the astronauts, but when the reality of the moon landing really hit, it was awesome.

On that historic Sunday in July we watched TV, laughing as Armstrong and Buzz Aldrin loped on the moon's surface. We walked out the front door and looked up at the Old Man and said, "It's really happening—there are a couple of guys walking around up there right now. Amazing."

Astounding as man's trip to the moon is, there is another trip which many men, women, and children will take some day which will leave the rest of the world gasping. Those who remain on earth at that time will use every invention of the human mind to explain the sudden disappearance of millions of people.

Reporters who wrote the historic story of Apollo 11 told how the astronauts collected rocks which may reveal the oldest secrets of the solar system. Those who are alive to tell the story of "Project Disappearance" will try in vain to describe the happening which will verify the oldest secrets of God's words.

What Will They Say?

"There I was, driving down the freeway and all of a sudden the place went crazy . . . cars going in all directions . . . and not one of them had a driver. I mean it was wild! I think we've got an invasion from outer space!"

"It was the last quarter of the championship game and the other side was ahead. Our boys had the ball. We made a touchdown and tied it up. The crowd went crazy. Only one minute to go and they fumbled—our quarterback recovered—he was about a yard from the goal when—zap—no more quarterback—completely gone, just like that!"

"It was puzzling—very puzzling. I was teaching my course in the Philosophy of Religion when all of a sudden three of my students vanished. They simply vanished! They were quite argumentative—always trying to prove their point from the Bible. No great loss to the class. However, I do find this disappearance very difficult to explain."

"As an official spokesman for the United Nations I wish to inform all peace-loving people of the world that we are making every human effort to assist those nations whose leaders have disappeared. We have issued a general declaration of condemnation in the General Assembly concerning these heads of state. Their irresponsibility is shocking."

"My dear friends in the congregation. Bless you for coming to church today. I know that many of you have lost loved ones in this unusual disappearance of so many people. However, I believe that God's judgment has come upon them for their continued dissension and quarreling with the great advances of the church in our century. Now that the reactionaries are removed, we can progress toward our great and glorious goal of uniting all mankind into a brotherhood of reconciliation and understanding."

"You really want to know what I think? I think all that talk about the Rapture and going to meet Jesus Christ in the air wasn't crazy after all. I don't know about you, brother, but I'm going to find myself a Bible and read all those verses my wife underlined. I wouldn't listen to her while she was here, and now she's—I don't know where she is."

Rapture—What Rapture?

Christians have a tendency sometimes to toss out words which have no meaning to the non-Christian. Sometimes misunderstood terms provide the red flag an unbeliever needs to turn him from the simple truth of God's Word. "Rapture" may be one of those words. It is not found in the Bible, so there is no need to race for your concordance, if you have one. There are some Christians who do not use the word, but prefer "translation" instead.

The word "rapture" means to snatch away or take out. But whether we call this event "the Rapture" or the "translation" makes no difference—the important thing is that it will happen.

It will happen!

Someday, a day that only God knows, Jesus Christ is coming to take away all those who believe in Him. He is coming to meet all true believers in the air. Without benefit of science, space suits, or interplanetary rockets, there will be those who will be transported into a glorious place more beautiful, more awesome, than we can possibly comprehend. Earth and all its thrills, excitement, and pleasures will be nothing in contrast to this great event.

It will be the living end. The ultimate trip.

If you are shaking your heads over this right now, please remember how many "impossibles" you have said in your lifetime—or how many "impossibles" men throughout the ages have said to many things God has revealed through His spokesmen. And yet they were possible, because nothing is impossible for God.

We have been examining the push of world events which the prophets foretold would lead the way to the seven-year countdown before the return of Jesus Christ to earth. The big question is, will you be here during this seven-year countdown? Will you be here during the time of the Tribulation when the Antichrist and the False Prophet are in charge for a time? Will you be here when the world is plagued by mankind's darkest days?

It may come as a surprise to you, but the decision

concerning your presence during this last seven-year period in history is entirely up to you.

God's Word tells us that there will be one generation of believers who will never know death. These believers will be removed from the earth before the Great Tribulation—before that period of the most ghastly pestilence, bloodshed, and starvation the world has ever known.

Examine the prophecies of this mysterious happening—of the "Rapture." Here is the real hope for the Christian, the "blessed hope" for true believers (Titus 2:13–15).

As we see the circumstances which are coming on the world, this hope gets more blessed all the time. This is the reason we are optimistic about the future. This is the reason that in spite of the headlines, in spite of crisis after crisis in America and throughout the world, in spite of the dark days which will strike terror into the hearts of many, every Christian has the right to be optimistic!

You may be thinking now, "Count me out. I like it right here and I have a lot of plans for my future."

Exactly. This is what we are talking about—your plans for the future. In fact, this is what Christ was talking about when He said: "Let not your heart be troubled; believe in God, believe also in Me. In My Father's house are many dwelling places; if it were not so, I would have told you; for I go to prepare a place for you. And if I go and prepare a place for you, I will come again, and receive you to Myself; that where I am, there you may be also" (John 14:1–3 NASB).

According to all the Scriptures we are told that the place He is preparing for us will be utterly fantastic. Eternal life will surpass the greatest pleasures we have known on earth.

I Tell You a Mystery

To avoid confusion, we will refer to the event when the church (those who believe in Jesus Christ as Savior) will meet Christ in the air as the Rapture. If you have grown in a school of Christian thought that uses the "translation," simply substitute this word when you are reading.

In I Corinthians 15:50 important things are revealed about the Rapture. We are told that Christians cannot inherit the Kingdom of God in the type of bodies we now have—that is, in bodies of flesh and blood.

However, according to the gospels and the Old Testament, there will be certain people who will inherit for a time the Kingdom of God in their mortal bodies. This is the Kingdom that Christ will establish after He returns to earth. This does not contradict the previous statement. The Bible is speaking of two separate events.

The distinction between God's dealing with the church and His dealing with another group of believers who are largely gathered around Israel is very important. Revelation 20 and Matthew 25 speak of the time when Jesus will return to the earth and separate the believers from the nonbelievers. For us, as believers, our hope is different from Israel's. This will be clear when we distinguish between the second advent, or the second coming of Christ, and the Rapture.

We are told that we cannot enter the Kingdom of God until we are changed from the type of body we now have into a new model. Then we have this fascinating verse: "Behold, I tell you a mystery; we shall not all sleep, but we shall all be changed" (I Corinthians 15:51 NASB).

The word "mystery" in the original Greek means something which has not been revealed before, but is now being revealed to those who are initiated. It was from this word that the concept of Greek fraternities came—everyone who has been in a fraternity or sorority knows there are certain secrets which are not disclosed until after initiation.

To draw the analogy, every believer in Jesus Christ is initiated into Christ's fraternity. Then, and only then, can he understand some of the secrets of God. These secrets the rest of the world will not accept as those who believe in Christ will accept them and understand them.

What is the secret that had not been revealed anywhere in the Scriptures before Paul wrote this letter to the Corinthians? Here is where the mystery gets exciting.

It says, "We shall not all sleep." Sleep is the word for Christian death. "Sleep" does not mean that your soul, your consciousness sleeps. There are some who believe that when you die your soul, your personality, the real you goes into some strange limbo. However, we are told that the moment a believer breathes his last breath and dies his soul goes immediately to be with Christ—to be face to face with the Lord (II Corinthians 5:1–10; Philippians 1:21–23).

So what does sleep? Your body. The body that disintegrates, Christ will raise into a body which can never see corruption again. "For our citizenship is in heaven, from which also we eagerly wait for a Savior, the Lord Jesus Christ; who will transform the body of our humble state into conformity with the body of His glory, by the exertion of the power that He has even to subject all things to Himself" (Philippians 3:20, 21 NASB).

What about the mystery? The mystery has to do with the believers who will be *alive* when Christ comes for them. "In a moment, in the twinkling of an eye, at the last trumpet; for the trumpet will sound, and the dead will be raised imperishable, and we shall be changed" (I Corinthians 15:52 NASB).

Words are fascinating—and the basis for understanding. The word translated "moment" is the Greek word *atomos,* which is the word from which we get "atom." In the Greek it means that which is indivisible, in other words, it will happen so quickly, in a flash of time which is so short it can't be divided. At that point those who are alive will be brought into the presence of the Lord.

This will take place "at the last trumpet," which refers to something which was the practice of God in the Old Testament. When the Israelites were on their march from Egypt over to the land of Palestine, every morning before they started on their journey they would have seven trumpets blow—to prepare to break camp, fold up their tents, etc. When the seventh trumpet, which was the last trumpet, sounded, this meant—move out!

The idea in this passage is that when God has the last trumpet blow it means He will move out all the Christians—and at that point we shall be changed.

What's in a word? "Changed" means to be changed in essence, but not to be completely changed in appearance. This strengthens the truth which is spoken of in other places that in eternity we are going to recognize people we knew here on earth. If you're not too satisfied with the face or body you now have, you will have a glorious new body. However, you will be recognizable, just as you will recognize others.

We won't have to eat to be sustained, but the Scripture says we can eat if we want to—and enjoy it. For those who have a weight problem, that sounds rather heavenly in itself. Our eternal bodies will not be subject to aging, or pain, or decay.

Just think how excited a woman can get about a new wardrobe. How much more excited we should be about acquiring a new body!

When the Scripture says, "the dead will be raised imperishable" and "for this perishable must put on the imperishable," it refers to the Christians who have died physically. They will be resurrected to meet Christ in the air.

However, when it says, "this mortal must put on immortality," it is referring to those who are alive at the coming of Christ. That's the mystery, the Rapture or translation. That is the hope that Paul offered for the generation which will be alive when Christ returns.

Who Goes First?

The Thessalonians were evidently worried about something that might be concerning you also. They wondered if those who had died and would be resurrected when Christ returned might be in some separate part of God's Kingdom. No Christian would want to miss seeing their loved ones throughout eternity.

However, the apostle Paul assured them that God's plan was perfect: those who had "fallen asleep in Jesus," or the Christians who had died, will join the Lord first. Then the Christians who are alive at that time will be caught up "together with them in the clouds" to meet the Lord in the air (I Thessalonians 4:13–18).

What a great reunion that will be!

The world will not know what has happened, because it occurs in an atom of time.

Debating Another Mystery

Christians sometimes have a theological debate about whether the Rapture occurs at the same time as the second coming of Christ or whether it takes place before the second coming, even before the Tribulation.

It is only fair to sincere Christians who differ about this time element for us to develop the reasons why we believe the Bible distinguishes between the Rapture and the second coming of Christ and why they do not occur simultaneously.

First, there is a great distinction between God's purpose for the nation of Israel and His purpose for the church, which is His main program today. The church is composed of both Gentiles and Jews. We are now living during the church age and the responsibility for evangelizing the world rests upon the church. We should re-emphasize here that we are speaking of the true meaning of the church, which is the body of believers in Jesus Christ.

In the Old Testament evangelizing was the task of the Jew. Of course he seldom fulfilled that obligation, which is one of his great failures. But the Scriptures give a vast distinction between God's dealing with the church and that time of Tribulation which seems to be a resumption of God's dealing with Israel. During the Tribulation the spotlight is on the Jew—in the Book of Revelation the Jew is responsible for evangelizing the world again (Revelation 7:1–4).

Another reason why we support the idea that the Rapture and the second coming are separate events is that the second coming is said to be visible to the whole earth (Revelation 1:7). However, in the Rapture, only the Christians see Him—it's a mystery, a secret. When the living believers are taken out, the world is going to be mystified.

Furthermore, we are told when Christ comes at the second coming it is at the height of a global war. Everyone will know that this is the great war predicted by the prophets. There will be no doubt about it. But when Christ comes for the believers, it will not necessarily be at the time of a war.

More proof—when Christ comes to earth for the second time we are told in Matthew 25 that He will divide the believers from the unbelievers. Now if the Rapture were to take place at the same time as the second coming how could the believers and unbelievers be separated on earth? At the Rapture all the living believers will be caught up to join Him in the clouds.

Here is the chief reason why we believe the Rapture occurs before the Tribulation: the prophets have said that God will set up a Kingdom on earth over which the Messiah will rule. There will be mortal people in that Kingdom. If the Rapture took place at the same time as the second coming, there would be no mortals left who would be believers; therefore, there would be no one to go into the Kingdom and repopulate the earth.

We need to understand that during the seven-year Tribulation there will be people who will become believers at that time. In spite of persecution as described in the previous chapter, they will survive this terrible period of history and will be taken by Christ to reign with Him for 1000 years. This is the Kingdom which is God's prelude to eternity.

The Church Disappeared

The largest descriptive volume of the Tribulation is found in Revelation 6 through 19. Here is a fascinating revelation about Revelation. In the first five chapters of this book, the church is mentioned thirty times. In fact, in chapters 2 and 3, at the end of each letter to the churches, John says "let him hear what the Spirit saith unto the churches." This is repeated seven times. Then we have the beginning of the description of the Tribulation, and there is not one mention of the churches. The church is conspicuous by its absence. Why? Because the church will be in heaven at that time.

If you are a believer, chapters 4 and 5 of Revelation describe what you will be experiencing in heaven. Talk about mind expansion drugs! We are told we shall expand in understanding and comprehension beyond that of any earthbound genius.

Be Alert

When will the Rapture occur? We don't know. No one knows. But God knows. However, we believe that according to all the signs, we are in the general time of His coming. "But you, brethren, are not in darkness, that the day should overtake you like a thief" (I Thessalonians 5:4 NASB).

In other words, you shouldn't be surprised when Christ returns to take you with Him. Unfortunately, this does not refer to all believers. We may have to go over to some of them and say, "I told you so, friend." It will be a surprise because they don't study the prophetic word. What an exciting time they may have missed on earth! The study and understanding of prophecy is an experience we pray all Christians will have.

"We are not of the night, nor of darkness . . ." (I Thessalonians 5:5 NASB). Darkness refers to the persons who do not have Christ, who cannot understand these things at all.

If you have not accepted Christ in your life this chapter will probably sound like the biggest farce you've ever read.

". . . so then let us not sleep as others do, but let us be alert and sober" (I Thessalonians 5:6 NASB). The idea of sleep is that a person just doesn't know what is going on. He may rationalize along his merry way, not paying any attention to the indications that the world can't go on much longer in the way it is going. He will say, "Something will happen—science will pull something out of the hat." So he puts his faith in science.

If you know what the prophets have said, and if the spirit of God has spoken to you, then you should be alert.

There's nothing that remains to be fulfilled before Christ could catch up to be with Him.

What's Important?

Have you ever found an electric train, or a bedraggled doll that belonged to you as a child and remembered how terribly important it was to you years ago? When we meet Christ face to face we're going to look back on this life and see that the things we thought were important here were like the discarded toys of our childhood.

What a way to live! With optimism, with anticipation, with excitement. We should be living like persons who don't expect to be around much longer.

There is no defense in science against the weapons which can now destroy civilization.

ALBERT EINSTEIN

In the next war, none of us can count on having enough living to bury our dead.

J. ROBERT OPPENHEIMER

Mankind must put an end to war—or war will put an end to mankind.

JOHN F. KENNEDY, 1961

A war would be an irreversible and fatal occurrence. It would not be the end of difficulties but the end of civilization.

POPE PAUL VI

. . . you will be hearing of wars and rumors of wars —then there will be a great tribulation, such as has not occurred since the beginning of the world until now, nor ever shall. And unless those days had been cut short, no life would have been saved.

JESUS CHRIST, A.D. 33

12

WORLD WAR III

From the beginning of man's history he has sought peace, but war has been his chief legacy. Among certain people today the peace sign has become an outward expression of their revulsion against war. This desire is like the people of Jeremiah's day who said, ". . . peace, peace, when there is no peace" (Jeremiah 6:14).

The great men of our day warn us about the insanity of another major war. Many experts, however, feel that it is inescapable. Whether we agree with the conclusions or theories of some of these experts or not, it is important to know what they are saying.

Several years ago a number of Nobel prize-winning scientists from various countries prepared a document and sent it to the leaders of all the world powers. They warned: "Here, then, is the problem which we present to you, stark and dreadful and inescapable: shall we put an end to the human race or shall mankind renounce war? We appeal, as human beings to human beings; remember your humanity, and forget the rest. If you can do so, the way lies open to a new paradise; if you cannot, there lies before you the risk of universal death." [1]

Many of the scientists who were most responsible for the development of the H-bomb signed the document quoted above.

War has greatly increased in frequency and intensity in this century. It has kept pace with the acceleration of technological advances. Some people have undoubtedly become callous to the continuous fighting on our globe, but it is shocking to review the statistics since World War II:

"Since World War II there have been 12 limited wars in the world, 39 political assassinations, 48 personal revolts, 74 rebellions for independence, 162 social revolutions, either political, economic, racial or religious." [2]

Since these statistics were written there have been more major assassinations and several more revolts.

In spite of all the oratory and books that have been aimed at steering man away from another world conflict, all-out war continues to be an ever-impending possibility. Any one of the limited wars such as Viet Nam or the Middle East crisis could at any time strike the spark that ignites World War III.

Why is it that in spite of the terrible lessons learned from history about war and the terrifying predictions of a future war, man keeps playing on the precipice of complete destruction? Jesus predicted that man would not learn from the past nor heed the warnings of the future; man would ultimately plunge the whole world into a war so vast, so utterly destructive, that only the personal return of Jesus Christ Himself to stop it would prevent the total annihilation of all life.

Here is the further solemn prediction of Jesus Christ as He described the world situation that would be present at the moment of His return:

"And unless those days had been cut short [abruptly ended], no life would have been saved . . ." (Matthew 24:22 NASB).

Why Can't Man Live Without War?

We believe the answer to that question is important to consider before presenting the predicted path that man will take to the last great war which the Bible calls Armageddon.

Man cannot stop war because he will not accept the basic reason and cause for war—nor will he accept the cure for this basic cause. God says: "From whence come wars and fightings among you? come they not hence, even of your lusts that war in your members? Ye lust, and have not: ye kill, and desire to have, and cannot obtain: ye fight and war, yet ye have not . . ." (James 4:1, 2 KJV).

Inside of man there is a selfish, self-centered nature. This is the source of what God calls sin. Sin is basically self-centered seeking and striving—going our own way, with our backs turned on God. It is because of this selfish nature with which we were born that we cannot have consistent peace with ourselves, our family, our neighbor, or, on a broader scale, with other nations.

As one man has said, "What's wrong with the world?" And answered himself truthfully, "I'm wrong with the world."

God didn't intend for man to have this condition. Man was created originally to have fellowship with God. This fellowship is so vital that without it man is like a jet aircraft flying in a dense fog and suddenly losing all its instruments.

God warned man of the consequences when He gave man one simple prohibition. Man understood that to disobey God on this one command was to reject fellowship with God. In spite of all the evidences of God's love and veracity, man went his own independent way and lost fellowship with the only One who can give him the true fulfillment for which he was created.

Man has been turned in upon himself; he has become self-centered and discontented. No matter how much he gets of fame or wealth or power, he is not satisfied. Why? Because he cannot fill the vacuum once filled by God with any other thing. So he fights with himself, with his mate, with his family, and with other nations.

Solution

The only cure for war is to change the hearts of men. Jesus came into the world to bring men back to fellowship with God and consequently make that change in their hearts. The Bible promises, "For Christ also died for sins once for all, the just for the unjust, in order that He might bring us to God . . ." (I Peter 3:18 NASB).

Jesus took the just rap due our sins and died under its penalty so that God might forgive us and receive us back into fellowship. When this occurs God gives us a new heart that desires to follow Him and a love for our neighbor. He gives us a new dimension of life with which we can perceive and know Him. He puts His Spirit within us to live so that He motivates and empowers us to follow God's purpose for our lives.

We begin to experience peace of mind, a new stability, a new sense of purpose, an awareness of Christ's presence, and a healing of our personality that makes us a whole person. We find ourselves concerned about the interests of others. A new kind of love impels us to place others before ourselves. This cannot be done by any system of government, education, psychology, or outer environmental changes. It can be done only by a personal invitation of Christ into the heart and an acceptance of the gift of forgiveness which He gave His life to provide.

Right where you are, as you read this chapter, you can make this decision in the quiet of your heart.

General Douglas MacArthur was eternally right when he said on the deck of the battleship *Missouri* at the close of World War II: "We have had our last chance. If we will not devise some greater and more equitable system, ARMAGEDDON will be at the door. The problem basically is theological and involves a spiritual recrudescence and improvement of human character that will synchronize with our almost matchless advances in science, art, literature, and all material and cultural developments of the past 2000 years. It must be the spirit if we are to save the flesh."[3]

The sad prediction of the Bible is that mankind will not accept God's diagnosis or His cure. Therefore, they will seek to solve the problem themselves. Fear of war will grow until it prepares man to accept the Antichrist's solution for preventing war. Paul predicts the false hope the world will have in the Antichrist: "While they are saying, 'Peace and safety!' then destruction will come upon them suddenly like birth pangs upon a woman with child; and they shall not escape" (I Thessalonians 5:3 NASB).

Where Are We Going?

In the previous chapters we have shown the predicted powers that would arise shortly before the return of Jesus Christ and how these powers are simultaneously developing in current history. We have outlined how world conditions in this generation are launched into a countdown that will end in the final collapse of man's efforts to run the world without God.

In this chapter we will trace consecutively the predicted events that lead to the Armageddon campaign: the various sequence of battles, the particular powers who fight each other, and how in turn each is destroyed. The crucial prediction of the revived state of Israel's part in triggering Armageddon will also be shown.

The Fuse Was Primed

When the Jews re-established their nation in Palestine they created an unsolvable problem; they displaced Arabs who had dwelt in Palestine for several centuries. All the legal debates and logical dissertations that can be advanced will never change the basic state of hostility that exists between the Israelis and the Arabs.

The Jews will never be convinced that they should leave the land that God gave to their forefathers. They believe that they were robbed of their inalienable right to the land by the Romans. Centuries of persecution have taught them that there is no country in the world where they can be assured of continuing acceptance, much less safety. Remaining as a nation in Palestine is a matter of survival of the race for the Israeli. He feels that it is his only hope in a hostile world.

The Arabs are equally implacable in their unwillingness to accept the Israeli occupation of what they consider to be their land. It has become a matter of racial honor and sacred religious duty to drive out the Israelis.

Israel's Treaty With Hell

According to the Bible, the Middle East crisis will continue to escalate until it threatens the peace of the whole world. The focus of all nations will be upon this unsolvable and complex problem which keeps bringing the world to the precipice of a thermonuclear holocaust. This is apparently the first major problem that the incredible Roman leader will solve after taking over the ten-nation confederacy of European nations.

Some 2500 years ago the prophet Daniel said that a prince would come to power from the people who would destroy the city of Jerusalem and the second Temple (Daniel 9:27). The Romans under Titus did the destroying, so the coming prince would have to be someone out of the Roman culture. This Roman prince, as we described in "The Future Fuehrer," will come to power just before the return of Christ. He will make "a strong covenant" with the Israelis, guaranteeing their safety and protection. The word translated "strong covenant" has the idea of a treaty or mutual protection pact. The Israelis will then be permitted to reinstitute the sacrifice and offering aspect of the law of Moses. This demands that the Temple be rebuilt, because according to the law of Moses, sacrifices can be offered only in the Temple at Jerusalem. Apparently all this will be done under the protection of the Antichrist of Rome.

(P.S. The Arabs are not going to like this idea of rebuilding the Temple one bit.)

According to Daniel's prophetic chronology, the minute the Israeli leader and the Roman leader sign this pact, God starts His great timepiece which has seven allotted years left on it. This event marks the beginning of the period of Biblical history previously noted as the Tribulation.

Isaiah prophetically expressed warning to the Jews concerning this covenant when he declared: "Because you have said, 'We have made a covenant with death, and with Sheol we have an agreement; when the overwhelming scourge passes through it will not come to us; for we have made lies our refuge, and in falsehood we have taken shelter'; Then your covenant with death will be annulled, and your agreement with Sheol will not stand; when the overwhelming scourge passes through you will be beaten down by it" (Isaiah 28:15, 18).

It is through an ingenuous settlement of the Middle East problem that the Antichrist will make good his promise to bring peace to a world terrified of war. After this he will rapidly bring all nations under his control. The world will experience great hope and put its full trust in the genius of Rome. He will begin to bring in fantastic plans of economic prosperity, even to the underdeveloped countries. War will seem to be a curious game that men used to play. The world will be universally acclaiming the Dictator.

"Who is like the Dictator, and who is able to make war with him?"

After three and a half years of remarkable progress, the Antichrist will become worshiped for his brilliant statesmanship and the wonderful progress in the world. The believers in Christ will oppose his rule and be ruthlessly exposed. Publicly, they will not be able to buy, sell, or hold a job. They will be executed en masse as examples to those who would hinder the "brotherhood of man," because they will insist that Christ is the only lasting hope for man.

Riding upon the crest of public worship the Roman Dictator will go to Jerusalem and in the Temple proclaim himself to be God incarnate (II Thessalonians 2:4, Matthew 24:15). As mentioned, this will be the great warning sign to the believers of that day that Armageddon is about to begin. The residents of Israel who believe in Jesus will flee to the mountains and canyons of Petra for divine protection, as promised (Matthew 24:16; Revelation 12:6, 14).

The Red Horse Unleashed

"And another, a red horse, went out; and to him who sat on it, it was granted to take peace from the earth, and that men should slay one another; and a great sword was given to him" (Revelation 6:4 NASB).

Almost immediately after the Antichrist declares himself to be God, God releases the dreaded second of the four horsemen of the Apocalypse. This is a figure of the unleashing of war upon the earth.

That beautiful balance of power established by the Antichrist is suddenly ruptured. God begins to show man that the Antichrist's promises cannot stand. The thing which man feared most, an all-out war, now rushes upon him.

The Beginning of the End

"At the time of the end the king of the south shall attack him (Israeli leader)" (Daniel 11:40a).

We have identified the characters of this passage. The Arab-African confederacy headed by Egypt (King of the South) launches an invasion of Israel. This fatal mistake spells their doom and begins the Armageddon campaign. ". . . but the king of the north shall rush upon him (the Israeli leader) like a whirlwind, with chariots (mechanized army) and horsemen (cavalry), and with many ships" (Daniel 11:40b).

Chart one shows the movement of troops.

Russia and her allies use this occasion to launch an invasion of the Middle East, which Russia has longed to do since the Napoleonic wars. Ezekiel 38 describes the development of this great Russian force, and its plan to attack Israel.

Twenty-six centuries ago Ezekiel described the plot of the Russian leaders against revived Israel: "Thus says the Lord GOD; On that day thoughts will come into your mind, and you will devise an evil scheme and say, 'I will go up against the land of unwalled villages; I will fall upon the quiet people who dwell securely, all of them dwelling without walls (i.e., fortifications), and having no bars or gates'; to seize spoil and carry off plunder; to assail the waste places which are now inhabited, and the people who were gathered from the nations, who have gotten cattle and goods, who dwell at the center of the earth" (Ezekiel 38:10–12).

The Russians will make a great tactical blunder by invading Israel. They will construe the defenseless posture of Israel, who will be trusting in the Antichrist's protection, as an opportunity to finally conquer the great land bridge of the Middle East. They will be motivated by the great material wealth of the restored nation of Israel. Their purpose is thus revealed by Ezekiel ". . . to seize spoil and carry off plunder. . . ."

The wealth of Israel is also predicted: ". . . the people who were gathered from the nations, who have gotten cattle and goods."

True Bible scholars have recognized that some day there would be a vast concentration of wealth in Israel. Harry Rimmer wrote in 1940, when the land was scrub brush in comparison to what it is today: ". . . The development of the resources of that land has only commenced. Ten years of uninterrupted industry there will make Palestine the richest concentration of treasure this world has so far witnessed. Even five years of unbroken application to the cultivation of her natural resources would make Palestine the envy of the world, and a land very well worth robbing, indeed." [4]

The prophetic indication is that Israel will become one of the most prosperous nations on earth during the reign of the Antichrist. It is also said to be, ". . . at the center of the earth." Geographically, this is approximately true, but more seems to be meant here than physical location. Israel will become a cultural, religious, and economic world center, especially at Jerusalem. The value of the mineral deposits in the Dead Sea

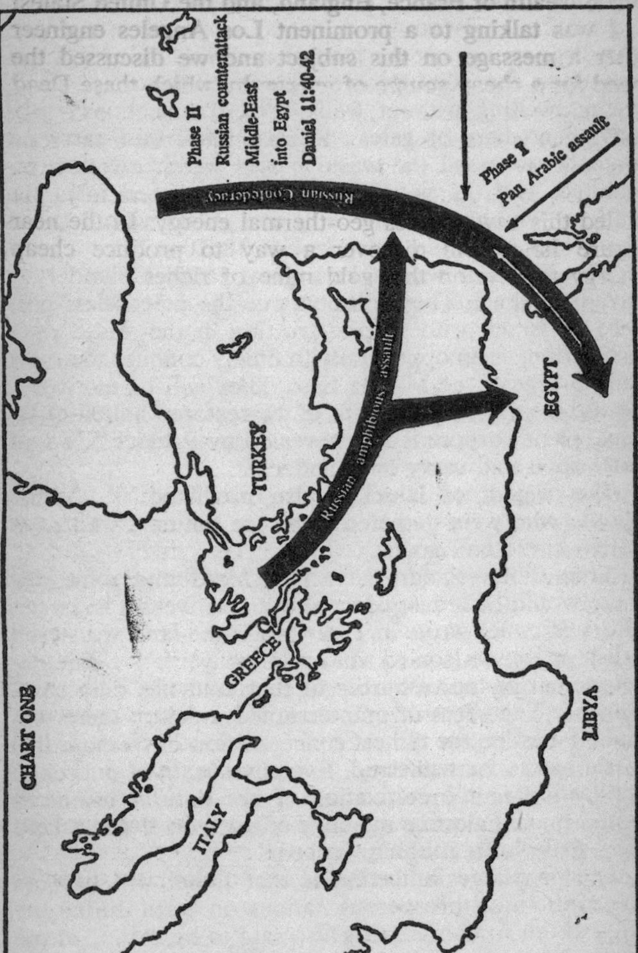
Phase II
Russian counterattack
Middle-East
into Egypt
Daniel 11:40-42
Russian Confederacy
Phase I
Pan Arabic assault
Russian amphibious assault
TURKEY
GREECE
ITALY
EGYPT
LIBYA
CHART ONE

alone has been estimated at one trillion, two hundred and seventy billion dollars. This is more than the combined wealth of France, England, and the United States!

I was talking to a prominent Los Angeles engineer after a message on this subject and we discussed the need for a cheap source of energy by which these Dead Sea minerals could be refined. He said that he is certain that there is enough steam trapped under the numerous faults in the earth around Israel to provide power to run turbines to produce electricity very economically. He called this new process geo-thermal energy. In the near future Israel will discover a way to produce cheap energy to develop this gold mine of riches.

One of the chief minerals in the Dead Sea is potash, which is a potent fertilizer. When the population explosion begins to bring famine, potash will become extremely valuable for food production.

It is strategic wealth of this sort that will cause the Russian bloc to look for an opportunity to invade and conquer Israel, according to Ezekiel.

The strategic military importance of "the land bridge of the Middle East" which begins in the north with the Bosphorus (the narrow channel which divides Turkey connecting the Black Sea with the Mediterranean), and extends southward to Egypt, has been established by centuries of military conflicts. The one who would control Europe, Asia, and Africa must control this area, which literally connects the three continents. This must figure into the Russian motivation for future conquest. Since Israel lies in the center of this land bridge, it has been made a battleground innumerable times.

This writer is indebted to Col. R. B. Thieme, Jr., a man who is uniquely qualified to explain this passage, since he is both a scholar in military history and in the original languages of the Bible, for pointing out the military aspects of this war.

The Classic Double-cross

When the Russians invade the Middle East with amphibious and mechanized land forces, they will make a "blitzkrieg" type of offensive through the area. As Daniel saw it centuries ago: ". . . and he [Russians] shall come into countries [of the Middle-East] and shall overflow and pass through. He will come into the glorious land [Israel]. And tens of thousands shall fall" (Daniel 11:40b, 41a).

Ezekiel describes the same invasion as follows: "Therefore, son of man, prophesy, and say to Gog [the Russian leader], Thus says the Lord GOD: On that day when my people Israel are dwelling securely, you will bestir yourself and come from your place out of the uttermost parts of the north [Daniel's king of the North], you and many peoples with you [i.e., the European iron curtain countries], all of them riding on horses, a great host, a mighty army; you will come up against my people Israel, like a cloud covering the land. In the latter days, I will bring you against my land . . ." (Ezekiel 38:14–16).

As previously quoted the Russians will make both an amphibious and land invasion of Israel. The current build-up of Russian ships in the Mediterranean serves as another significant sign of the possible nearness of Armageddon. They now have more ships in the Mediterranean than the United States, according to several recent news releases. The amphibious landings will facilitate a rapid encirclement of the middle section of "the land bridge."

The might of the Red Army is predicted. It will sweep over the Arab countries as well as Israel in a rapid assault over to Egypt to secure the entire land bridge. It is at this point that Russia double-crosses the United Arab Republic leader, Egypt. After sweeping over tens of thousands of people Daniel says of the Red Army: "He shall stretch out his hand against the countries [i.e. Arab countries of the Middle East], and the land of Egypt shall not escape. He [Russian leader] shall become ruler of the treasures of gold and of silver, and all the precious things of Egypt; and the Libyans [African Arabs] and the Ethiopians [African blacks] shall follow in his train" (Daniel 11:42, 43).

As we saw in Chapters 5 and 6, this prediction indicates that the Russian bloc will double-cross the Arabs,

Egyptians, and Africans, and for a short while conquer the Middle East. At this time, with the main Russian force in Egypt, the commander will hear alarming news: "But rumors from the east [the Orient mobilizing] and from the north [the Western Europeans mobilizing] shall alarm and hasten him. And he shall go forth with great fury to destroy and utterly to sweep away many" (Daniel 11:44 Amplified).

As shown in Chart two, the Russian force will retrace its steps from Egypt to consolidate for a counter-attack in Israel. The Russians will be alarmed at the news of the Roman Dictator mobilizing forces around the world to put down this breach of peace. Apparently it will surprise the Russian leader who underestimated the revived Roman Empire's will to fight.

It is conjecture on this writer's part, but it appears that the Oriental powers, headed by Red China, will be permitted to mobilize its vast army by the Roman Dictator, thinking that they would be loyal to him against Russia. However, the Orientals will eventually double-cross him, and move a 200 million man army against the Antichrist, as we have seen in Chapter 7.

This Russian double-cross of the Arabs is predictable by any astute observer of the Middle East situation today. It is obvious that the Russians are playing games with the Arabs in order to accomplish the old Russian dream of year-round seaports and oil supplies. The Arab leaders think that they can accept Russian loans and supplies without strings, but there are steel cables of conditions behind every Russian ruble given in aid and they are used eventually to pull a country behind the iron curtain.

The Russian force will establish command headquarters on Mount Moriah or the Temple area in Jerusalem. Daniel pointed this out when he said: "And he shall pitch his palatial tents between the seas [Dead Sea and Mediterranean Sea] and the glorious holy mount Zion; yet he shall come to his end with none to help him" (Daniel 11:45 Amplified).

As the Russian commander confidently prepares to meet the forces of the revived Roman Empire in Pales-

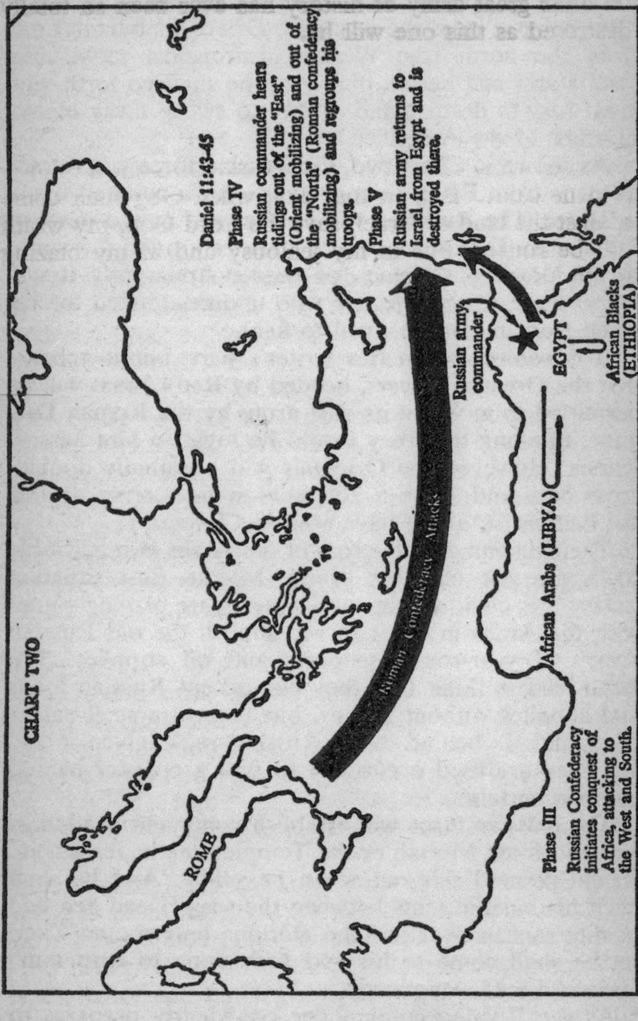
CHART TWO
Daniel 11:43-45
Phase IV
Russian commander hears tidings out of the "East" (Orient mobilizing) and out of the "North" (Roman confederacy mobilizing) and regroups his troops.
Phase V
Russian army returns to Israel from Egypt and is destroyed there.
Russian army commander
EGYPT
African Blacks (ETHIOPIA)
African Arabs (LIBYA)
Roman Confederacy Attacks
ROME
Phase III
Russian Confederacy initiates conquest of Africa, attacking to the West and South.

tine, he seeks to utterly destroy the Jewish people. This is the apparent meaning of verses 44 and 45. Perhaps no other great army of history has ever been so totally destroyed as this one will be.

The Russian Waterloo

Ezekiel sounded the fatal collapse of the Red Army centuries ago when inspired by the Spirit of the living God he said: "But on that day, when Gog shall come against the land of Israel, says the Lord GOD, my wrath will be roused. For in my jealousy and in my blazing wrath I declare, On that day there shall be a great shaking in the land of Israel; the fish of the sea, and the birds of the air, and the beasts of the field, and all creeping things that creep on the ground, and all the men that are upon the face of the earth, shall quake at my presence, and the mountains shall be thrown down, and the cliffs shall fall, and every wall shall tumble to the ground. I will summon every kind of terror against Gog, says the Lord GOD; every man's sword will be against his brother. With pestilence and bloodshed I will enter into judgment with him; and I will rain upon him and his hordes and the many peoples that are with him, torrential rains and hailstones, fire and brimstone" (Ezekiel 38:18–22).

". . . then I will strike your bow from your left hand, and will make your arrows drop out of your right hand. You shall fall upon the mountains of Israel, you and all your hordes and the people that are with you; I will give you to birds of prey of every sort and to the wild beasts to be devoured. You shall fall in the open field; for I have spoken, says the Lord GOD" (Ezekiel 39:3–5).

The description of torrents of fire and brimstone raining down upon the Red Army, coupled with an unprecedented shaking of the land of Israel could well be describing the use of tactical nuclear weapons against them by the Romans. It explicitly says that this force would fall "in the open field," so apparently this position enables the use of nuclear weapons.

God consigns this whole barbarous army, which will seek to annihilate the Jewish race, to an utter and complete decimation. Ezekiel speaks of the Russians and ". . . all your hosts and the peoples who are with you . . ." being destroyed in Israel.

Nuclear Exchange Begins

A fearful thing is predicted as occurring at the time of the Red Army's destruction. God says, "I will send fire on Magog [Russia] and upon those who dwell securely [false hope] in the coastlands [various continents] . . ." (Ezekiel 39:6 Amplified). According to this, Russia, as well as many countries who thought they were secure under the Antichrist's protection, will have fire fall upon them. Once again, this could be a direct judgment from God, or God could allow the various countries to launch a nuclear exchange of ballistic missiles upon each other.

What About the United States?

The United States may be aligned with the Western forces headed by the ten-nation Revived Roman Empire of Europe. It is clear that the U.S. cannot be the leader of the West in the future. It is quite possible that Ezekiel was referring to the U.S. in part when he said: "I will send fire—upon those who dwell securely in the coastlands. . . ."

The word translated "coastlands" or "isle" in the Hebrew is *ai*. It was used by the ancients in the sense of "continents" today. It designated the great Gentile civilizations across the seas which were usually settled most densely along the coastlands. The idea here is that the Gentile nations on distant continents would all experience the impact of sudden torrents of fire raining down upon them. This can include prophetically the populated continents and islands of the Western hemisphere as well as the Far East. It pictures cataclysmic events which affect the whole inhabited earth.

The Greatest Battle of All Time

With the United Arab and African armies neutralized by the Russian invasion, and the consequent complete annihilation of the Russian forces and their homeland, we have only two great spheres of power left to fight the final climactic battle of Armageddon: the combined forces of the Western civilization united under the leadership of the Roman Dictator and the vast hordes of the Orient probably united under the Red Chinese war machine.

The Mobilization

There may be a short lapse of hostility in the world after the destruction of Russia and allies while the armies mobilize. The apostle John predicts the mobilization of the Oriental power as mentioned in Chapter 7. "And the sixth angel poured out his bowl upon the great river, the Euphrates; and its water was dried up, that the way might be prepared for the kings from the east" (Revelation 16:12 NASB).

The Chinese, as previously discussed, will be the most logical leader of this great army. They will use the chaos caused by this conflict in the Middle East to march against the Roman Dictator in a challenge for world control.

As this incredible Oriental army of 200 million soldiers marches to the eastern banks of the Euphrates, the Roman Dictator will begin to prepare his armies to meet them for the showdown in the Middle East. The apostle John predicts the supernatural power that the Antichrist and the False Prophet will use to gather all remaining nations together to confront the Oriental armies: "And I saw coming out of the mouth of the dragon [Satan] and out of the mouth of the beast [Roman Dictator] and out of the mouth of the false prophet, three unclean spirits like frogs; for they are spirits of demons, performing signs, which go out to the kings of the whole world, to gather them together for war of the great day of God, the Almighty. And they gathered them together to the place which in Hebrew is called Armageddon" (Revelation 16:13, 14, 16 NASB).

Several things must be noted from this passage. First, this whole sequence of events constitutes the final judgments of God against the Christ-rejecting world. They are called in the Book of Revelation the seven bowl or vial judgments and they occur just before and during the visible return of Jesus Christ to the earth. Secondly, the passage shows that the Roman Dictator and his sensational religious cohort, the False Prophet, will speak a satanically originated message to the non-Oriental world—a message energized with great demonic power. They will persuade the nations of the whole world (i.e., not aligned with China) that they should send armies to the land of Palestine to destroy the last great warlike force on earth. They will probably promise an age of everlasting peace after the troublesome Communist forces of the Orient are destroyed. Thirdly, the passage indicates that "leaders of the whole world" will send armies to the Middle East to fight under the Antichrist's command against "the kings of the east." Such countries as Western Europe, the United States, Canada, South America, and Australia undoubtedly will be represented.

Fourthly, these armies will be assembled and deployed for battle in the place called "Armageddon" or "Harmageddon."

What and Where Is Armageddon?

Armageddon is a byword used through the centuries to depict the horrors of war. Dr. Seiss sums up its true significance as follows: "Harmageddon (Armageddon) means the Mount of Megiddo, which has also given its name to the great plain of Jezreel which belts across the middle of the Holy Land, from the Mediterranean to the Jordan. The name is from a Hebrew root which means to cut off, to slay; and a place of slaughter has Megiddo ever been." [5]

In Biblical history countless bloody battles were fought in this area. Napoleon is reported to have stood upon the hill of Megiddo and recalled this prophecy as he looked over the valley and said, ". . . all the armies of the world could maneuver for battle here." In the Old Testament book of Joel this valley was called the "valley of Jehoshaphat."

Today this valley's entrance has the port of Haifa at its Western end. This is one of the most accessible areas in Palestine for amphibious landing of troops. It also affords a great area for troop assembly, equipment, and organization. Some troops will doubtlessly be airlifted in as well, and this large valley is suited for that, too.

The Valley of Decision

Some twenty-seven centuries ago the prophet Joel focused upon this same scene and said: "Proclaim this among the nations: Prepare war, stir up the mighty men. Let all the men of war draw near, let them come up. Beat your plowshares into swords, and your pruning hooks into spears; let the weak say, 'I am a warrior.' Hasten and come, all you nations round about, gather yourselves there. Bring down thy warriors, O LORD. Let the nations bestir themselves, and come up to the valley of Jehoshaphat; for there I will sit to judge all the nations round about. Put in the sickle, for the harvest is ripe. Go in, tread, for the wine press is full. The vats overflow, for their wickedness is great. Multitudes, multitudes, in the valley of decision! For the day of the LORD is near in the valley of decision" (Joel 3:9–14).

Joel reveals that it is in this very place that the Messiah will destroy the armies of the world and establish his kingdom of true peace and everlasting happiness. He also confirms the world-wide assemblage of armies there.

It is extremely important to note the accuracy of Bible prophecy in relation to this last conflict. In this day of H-bombs and super weapons, it seems incredible that there could ever be another great land war fought by basically conventional means, yet the Chinese believe that with a vastly superior numerical force, they can absorb devastating losses and still win a war. They also believe that all war is still determined on the ground by land forces.

Another interesting fact is that a force of 200 million soldiers could not be transported by China and its foreseeable allies by air or sea. They do not possess the industry to produce a transportation system for such an army. This necessitates the movement of troops across the land as is indicated in Revelation 16:12.

India has revealed recently an important development. It is reported that 12,000 Chinese soldiers are at work inside Pakistan-held Kashmir on the road which would give Chinese troops in Tibet a shortcut to the subcontinent. India calls the Chinese road-building activities a "threat to peace in Asia." It was said that "a flurry of road-building throughout the Himalayas is taking on increasing strategic importance." [6]

When this road is completed, it will make possible the rapid movement of millions of Chinese troops into the Middle East. It will literally pave the way for John's prophecy to be fulfilled: ". . . unless those days had been cut short, no life would have been saved . . ." (Matthew 24:22 NASB).

So here it is—the last great conflict. After the Antichrist assembles the forces of the rest of the whole world together, they meet the onrushing charge of the kings of the East in a battle line which will extend throughout Israel with the vortex centered at the Valley of Megiddo.

According to Zechariah, terrible fighting will center around the city of Jerusalem (Zechariah 12:2, 3; 14:1, 2).

Isaiah speaks of a frightful carnage taking place south of the Dead Sea in ancient Edom (Isaiah 63:1–4).

The apostle John predicts that so many people will be slaughtered in the conflict that blood will stand to the horses' bridles for a total distance of 200 miles northward and southward of Jerusalem (Revelation 14:20).

It seems incredible! The human mind cannot conceive of such inhumanity of man to man, yet God will allow man's nature to fully display itself in that day. No wonder Jesus said: ". . . for then there will be a great tribulation, such as has not occurred since the beginning of the world until now, nor ever shall" (Matthew 24:21 NASB).

World-wide Destruction

The conflict will not be limited to the Middle East. The apostle John warns that when these two great forces meet in battle the greatest shock wave ever to hit the earth will occur. Whether by natural force of an earthquake or by some super weapon isn't clear. John says that all the cities of the nations will be destroyed (Revelation 16:19).

Imagine, cities like London, Paris, Tokyo, New York, Los Angeles, Chicago—obliterated! John says that the Eastern force alone will wipe out a third of the earth's population (Revelation 9:15–18).

He also predicts that entire islands and mountains would be blown off the map. It seems to indicate an all-out attack of ballistic missiles upon the great metropolitan areas of the world.

Prophecy indicates that U.S. Representative John Rhodes was right when he said concerning the danger of Red China and nuclear weapons, ". . . anyone who expects restraint from Red China ignores history and imperils future generations of Americans." [7]

Isaiah predicts in chapter 24 concerning this time: "Behold, the Lord will lay waste the earth and make it desolate, and he will twist its surface and scatter its inhabitants."

"The earth lies polluted under its inhabitants." (Perhaps this refers in part to water and air pollution.)

"Therefore, a curse devours the earth, and its inhabitants suffer for their guilt; therefore the inhabitants of the earth are scorched [burned], and few men are left" (verses 1, 5, 6).

In the same chapter Isaiah says: "The earth is utterly broken, the earth is rent asunder, the earth is violently shaken. The earth staggers like a drunken man, it sways like a hut . . ." (verses 19, 20).

All of these verses seem to indicate the unleashing of incredible weapons the world over.

A Bright Spot in the Gloom

As Armageddon begins with the invasion of Israel by the Arabs and the Russian confederacy, and their consequent swift destruction, the greatest period of Jewish conversion to their true Messiah will begin. Ezekiel predicts that the destruction of the great Russian invading force will have a supernatural element to it which will cause great numbers of Jews to see the hand of the Lord in it. Through the miraculous sign of the destruction of this enemy who sought to destroy all Jews they come to see the name of their true God and Messiah, Jesus Christ.

Ezekiel quotes God as saying: "I will send fire on Magog [Russia] and on those who dwell securely in the coastlands; and they shall know that I am the LORD. And my holy name I will make known in the midst of my people Israel; and I will not let my holy name be profaned any more; and the nations shall know that I am the LORD, the Holy One in Israel. Behold, it is coming and it will be brought about, says the Lord GOD. That is the day of which I have spoken" [i.e., in the prophecies] (Ezekiel 39:6–8).

Zechariah predicts that one-third of the Jews alive during this period will be converted to Christ and miraculously preserved.

"In the whole land, says the LORD, two thirds shall be cut off and perish, and one third shall be left alive. And I will put this third into the fire, and refine them as one refines silver, and test them as gold is tested. They will call on my name, and I will answer them. I will say, 'They are my people'; and they will say, 'The LORD is my God' " (Zechariah 13:8, 9).

The Greatest Moment

As the battle of Armageddon reaches its awful climax and it appears that all life will be destroyed on earth—in this very moment Jesus Christ will return and save man from self-extinction.

As history races toward this moment, are you afraid or looking with hope for deliverance? The answer should reveal to you your spiritual condition.

One way or another history continues in a certain acceleration toward the return of Christ. Are you ready?

The purposes of the United Nations are: to maintain international peace and security, and to that end: to take effective collective measures for the prevention and removal of threats to the peace.

U. N. CHARTER, 1945

These things I have spoken to you, that in Me you may have peace. In the world you have tribulation, but take courage; I have overcome the world.

JESUS CHRIST

13

THE MAIN EVENT

Written upon the cornerstone of the United Nations building is a quotation of part of a prophecy. It reads: ". . . they shall beat their swords into plowshares, and their spears into pruning hooks; nation shall not lift up sword against nation, neither shall they learn war any more" (Isaiah 2:4).

This is a noble thought and has been quoted often by men who seek peace for this troubled world. There is a problem in the phrase, however, and we believe this is why the United Nations will never bring a lasting peace to the world. This quotation has been taken out of context. The meaning of the passage speaks of the time when the Messiah would reign over the earth out of Jerusalem and judge between the nations in a visible, actual, and historic Kingdom of God on earth. The people of the earth will come to the Lord in that day and ask Him to teach them His ways. Knowledge of God will be universal (Isaiah 2:3). This is the era for which Jesus taught us to pray in the Lord's Prayer, ". . . Thy kingdom come, thy will be done on earth as it is in heaven" (Matthew 6:10).

Men today vainly seek after peace while they reject and shut out of their lives the Prince of Peace, Jesus

Christ. The name of Christ is not mentioned at the close of prayer in the United Nations. In fact, Jesus has been excluded from the premises. Man has shut out the only hope of peace, according to the Bible. The spirit of "antichrist" reigns in the governments of the world, for Christ is said not to be relevant to the problems we face.

Peace is available to the individual today as he invites Christ into his heart and allows Him to reign upon the throne of his life. But the Bible teaches that lasting peace will come to the world only after Christ returns and sits upon the throne of David in Jerusalem and establishes His historic kingdom on earth for a thousand years (Revelation 20:4–6).

The rulers of the world are told by many prophecies that God would send His king to rule over the earth and establish a reign of peace, righteousness, and justice in place of their godless, selfish, and violent rule. Jesus will return at a time of world-wide catastrophe, when man is on the brink of self-destruction. Men, for the most part, will have utterly rejected the true God and His Son, Jesus the Messiah, as predicted long ago by the psalmist: "The kings of the earth set themselves, and rulers take counsel together, against the LORD and his anointed [Christ], saying, 'Let us burst their bonds asunder, and cast their cords from us' " (Psalm 2:2, 3).

In spite of all that man will try to do to establish his rule and push God out of the world, God will establish His king, the Messiah Jesus, as the psalmist goes on to predict: "Yet have I set my king upon my holy hill of Zion" (Psalm 2:6 KJV).

Many so-called Christian leaders today do not believe that Jesus Christ will literally and physically make a personal return to the earth. Some teach that Christ returns spiritually when people accept Him and say that this is all that is meant by the various predictions of His return. Others teach that Jesus may return some day, but that it is irrelevant to study or to talk about it. The latter are worse than the former, for one out of every twenty-five verses in the New Testament is related to the second coming of Christ, and the survival of mankind as well as the fulfillment of hundreds of unconditional promises especially made to the believing remnant of the Jewish race are dependent on the second coming of Christ to this earth. As a matter of fact, in the Old Testament there were more than 300 prophecies regarding Christ's first coming (all of which were literally fulfilled), but more than 500 relating to His second coming. Many of these two different themes of prophecy were disclosed in the same sentence.

We have read that the apostle Peter clearly warned that in the days just prior to Christ's return false teachers would arise in the church and say: "Where is the promise of His (Jesus') coming? For ever since the fathers [apostles] fell asleep, all continues just as it was from the beginning of creation" (II Peter 3:4 NASB).

Characteristics of Christ's Return

Immediately after Jesus physically departed from the Mount of Olives and while His disciples were staring after Him with awe and amazement, the following promise was given: "And as they were gazing intently into the sky while He [Jesus] was departing, behold, two men in white clothing stood beside them; and they also said, 'Men of Galilee, why do you stand looking into the sky? This Jesus, who has been taken from you into heaven, will come in just the same way you have watched Him go into heaven' " (Acts 1:10, 11 NASB).

The word in the original translated "in just the same way" means "in exactly the same manner." Just as Jesus departed physically, visibly and personally from the earth, so He will return. Just as Jesus departed with clouds, so He will return.

His Return Will Be Visible, Personal, and Physical

The apostle John said: "Behold, He [Jesus] is coming with the clouds, and every eye will see Him, even those who pierced Him; and all the tribes of the earth will mourn over Him" (Revelation 1:7 NASB).

Zechariah predicted the same picture 500 years before Christ was born: ". . . and they [believing Israelites] shall look upon me [Jesus] whom they have pierced, and they shall mourn for him, as one mourneth for his only son . . ." (Zechariah 12:10 KJV).

To mourn over the one who was pierced necessitates that man recognizes Jesus who was crucified and rejected. This demands a dramatic personal and physical appearance.

Jesus promised under oath before the high priest at His trial: ". . . nevertheless I tell you, hereafter you shall see the Son of Man sitting at the right hand of Power [God], and coming on the clouds of heaven" (Matthew 26:64 NASB).

This statement was the official ground of His condemnation for blasphemy and the death sentence. Jesus dared to be the One who would fulfill two of the best-known prophecies concerning the Messiah's coming in glory to rule the earth. The first is from the Psalms, predicted before 1000 B.C.: "The LORD [God, the Father] said unto my Lord [God, the Son], Sit thou at my right hand, until I make thine enemies thy footstool" (Psalm 110:1).

The second is from Daniel, predicted about 550 B.C.: "I saw in the night visions, and behold, one like the Son of man came with the clouds of heaven, and came to the Ancient of days, and they brought him near before him. And there was given him dominion, and glory, and a kingdom, that all people, nations, and languages should serve him: his dominion is an everlasting dominion, which shall not pass away, and his kingdom that which shall not be destroyed" (Daniel 7:13, 14 KJV).

No wonder the Jewish supreme court (the Sanhedrin) went into orbit. When Jesus made such a fantastic claim as that in one terse sentence, they either had to fall down and worship Him or kill Him. They chose the latter.

His Coming Will Be Sudden and Startling

Jesus predicted the suddenness of His return as follows: "For just as lightning comes from the east, and flashes even to the west, so shall the coming of the Son of Man be" (Matthew 24:27 NASB).

And again He said: ". . . and then the sign of the

Son of Man will appear in the sky, and then all the tribes of the earth shall mourn, and they will see the Son of Man coming on the clouds of heaven with power and great glory" (Matthew 24:30 NASB).

Perhaps the "sign of the Son of Man" will be a gigantic celestial image of Jesus flashed upon the heavens for all to see. This would explain how all men suddenly recognize who He is and see the scars from His piercing at the cross.

His Coming Will Be With the Saints

It is significant to note that many references to Christ's return speak of His return accompanied with "the clouds of heaven." We believe that the clouds refer to the myriads of believers who return in white robes with Jesus. Believers are referred to as "a cloud of witnesses" in Hebrews 12:1. The clouds then would be all of the church age believers, you and I, returning in immortal glorified bodies, having been previously caught up to meet Christ in the air in "the ultimate trip," prior to the seven years of Tribulation on earth, and the resurrected saints of the Old Testament (Revelation 19:14).

The word "saint" means someone who is set apart as God's possession. It is used to designate all who have believed in Christ as Savior. This word is used many times to refer to those who will accompany Christ at His return.

As Zechariah referred to the Messiah's second coming he said: ". . . and the LORD my God, shall come, and all the saints with thee" (14:5 KJV).

The apostle John speaks of the apparel of the saints as they return with Christ: "And the armies which are in heaven, clothed in fine linen, white and clean, were following Him on white horses" (Revelation 19:14 NASB).

John explains the white linen robes: "And it was given to her [the church made up of all believers who have been caught up in the Rapture] to clothe herself in fine linen, bright and clean; for the fine linen is the righteous acts of the saints" (Revelation 19:8 NASB).

His Coming Will Be With Violent Judgment

When Jesus came the first time it was not to judge the world, but to save it. He came as the Lamb of God who gave His life to take away the sin of the world. The one thing that God has established for man to do is to believe in His Son as Savior. When Jesus returns the second time it will be as a lion to judge those who rejected the free gift of salvation from sin. Man will have completely demonstrated his worthiness of judgment.

According to Zechariah, "all nations will be gathered against Jerusalem to battle." The Jews who live in the area will be on the verge of annihilation when God gives them supernatural strength to fight. Then the Lord will go forth to fight for them and save them.

Jesus' feet will first touch the earth where they left the earth, on the Mount of Olives. The mountain will split in two with a great earthquake the instant that Jesus' foot touches it. The giant crevice which results will run east and west through the center of the mountain. It will go east to the north tip of the Dead Sea and west to the Mediterranean Sea (Zechariah 14).

It was reported to me that an oil company doing seismic studies of this area in quest of oil discovered a gigantic fault running east and west precisely through the center of the Mount of Olives. The fault is so severe that it could split at any time. It is awaiting "the foot."

Zechariah predicts a strange thing with regard to the ensuing split in the earth. The believing Jewish remnant in Jerusalem will rush into the crack instead of doing the natural thing of running from it. They will know this prophecy and realize that this great cavern has opened up for the Lord to protect them from the terrible devastation that He is about to pour out upon the godless armies all around. It will be used as a type of bomb shelter.

The nature of the forces which the Lord will unleash on that day against the armies gathered in the Middle East is described in Zechariah 14:12: "And this shall be the plague wherewith the LORD will smite all the people that have fought against Jerusalem; Their flesh shall consume away while they stand upon their feet, and their eyes shall consume away in their holes, and their tongue shall consume away in their mouth" (KJV).

A frightening picture, isn't it? Has it occurred to you that this is exactly what happens to those who are in a thermonuclear blast? It appears that this will be the case at the return of Christ.

His Return Will Be to Set Up God's Kingdom on Earth

After Christ destroys all ungodly kingdoms, Zechariah says, "The Lord shall be king over all the earth; in that day shall there be one Lord, and His name one."

Most ministers and religious leaders today reject even the possibility that Christ will establish an actual physical kingdom of God upon the earth. Many who believe in a personal return of Christ reject that He will establish a thousand year kingdom of God and rule mortals from the throne of David out of Jerusalem after His return.

The Latin word for "1000" is "millennium" and down through history the teaching concerning this earthly kingdom came to be known as the "millennial kingdom." Those who reject that Christ will establish a 1000 year kingdom after His return are known theologically as "amillennialists," meaning "no millennium." Those who believe that Christ will return and set up a 1000 year kingdom are called "premillennialists," meaning Christ returns first, then establishes the kingdom on earth.

There used to be a group called "postmillennialists." They believed that the Christians would root out the evil in the world, abolish godless rulers, and convert the world through ever increasing evangelism until they brought about the Kingdom of God on earth through their own efforts. Then after 1000 years of the institutional church reigning on earth with peace, equality, and righteousness, Christ would return and time would end. These people rejected much of the Scripture as being literal and believed in the inherent goodness of man. World War I greatly disheartened this group and World War II virtually wiped out this viewpoint. No self-respecting scholar who looks at the world conditions and the accelerating decline of Christian influence today is a "postmillennialist."

We are "premillennialists" in viewpoint. The real issue between the amillennial and the premillennial viewpoints is whether prophecy should be interpreted literally or allegorically. As it has been demonstrated many times in this book, all prophecy about past events has been fulfilled literally, particularly the predictions regarding the first coming of Christ. The words of prophecy were demonstrated as being literal, that is, having the normal meaning understood by the people of the time in which it was written. The words were not intended to be explained away by men who cannot believe what is clearly predicted.

The opponents of the premillennial view all agree grudgingly that if you interpret prophecy literally it does teach that Christ will set up a literal kingdom in time which will last in history a thousand years and then go into an eternal form which will never be destroyed.

To us the biggest issue is over the question, "Does God keep His promises?" For God unconditionally promised Abraham's descendants a literal world-wide kingdom over which they would rule through their Messiah who would reign upon King David's throne. The Jews who believe in the Messiah will also possess the land which is bordered on the east by the Euphrates River, and on the west by the Nile (Genesis 15:18–21).

It is promised that Jerusalem will be the spiritual center of the entire world and that all people of the earth will come annually to worship Jesus who will rule there (Zechariah 14:16–21; Isaiah 2:3; Micah 4:1–3). The Jewish believing remnant will be the spiritual leaders of the world and teach all nations the ways of the Lord (Zechariah 8:20–23; Isaiah 66:23).

Paradise Restored

God's kingdom will be characterized by peace and equity, and by universal spirituality and knowledge of the Lord. Even the animals and reptiles will lose their ferocity and no longer be carnivorous. All men will have plenty and be secure. There will be a chicken in every pot and no one will steal it! The Great Society which human rulers throughout the centuries have promised, but never produced, will at last be realized under Christ's rule. The meek and not the arrogant will inherit the earth (Isaiah 11).

Prelude to Eternity

As it was mentioned in Chapters eight and nine, Daniel predicted the four great world ruling kingdoms that man would set up during the time from the sixth century B.C. until the coming of the Messiah. We noted these four human empires as Babylon, Media-Persia, Greece, and Rome, with its revived form in the last days. The fifth world kingdom, which according to Daniel will conquer the revived form of the Roman Empire, is the Messianic kingdom (Daniel 7:13–27).

This kingdom will begin in time with mortal subjects (Revelation 20:4–6), last 1000 years, and at the end of that time some of the children of the believers who started in the kingdom will apparently prove to be unbelievers and start a rebellion against Christ and His rule. Christ will bring swift judgment upon them before the rebellion reaches the actual fighting stage (Revelation 20:7–10).

After this event there will be no more human history with mortal men. All unbelievers, it seems, will be judged in the crushing of the last rebellion which is led by God's old adversary, Satan. Satan will have been bound for a thousand years, but is released momentarily so that he could reveal the rebellion in the unbelieving hearts of those who rejected Christ as Savior (Revelation 20:7, 8). All who remain as mortals will be changed into immortality at this point, and the Kingdom of God will not cease, but simply change form and be re-established in a new heaven and a new earth (Revelation 21).

The sequence is clear in the last chapters of Revelation. First there is the return of Christ at the climax of the greatest war of all time. Second, Christ separates the surviving believers from the surviving unbelievers; the unbelievers will be judged and cast off the earth (Revelation 20:1–6; cf. Matthew 25:41–46). Third, Christ establishes the millennial kingdom and the surviving believers go into it as mortals and repopulate the earth (Revelation 20:11–15; cf. Matthew 25:31–40). Fourth, at the end of a thousand years the unbelieving children rebel, Christ judges them, then He completely changes the old heaven and earth and creates a new one (Revelation 21; Isaiah 65:17; II Peter 3:8–13). This is the ultimate destiny of all persons who are redeemed by Christ.

How many times have we wondered what heaven will be like? According to such passages as Revelation 21 and 22, heaven is a real and breath-taking place. We will not wander about as disembodied spirits, playing harps throughout an ethereal expanse. We shall live forever in the presence of God, fellow heirs with Christ, as kings and priests forever, with no more sorrows or tears. We shall know an ecstatic, endless joy surrounded by an earth and heaven of indescribable beauty. If you can think of the most beautiful place you have ever been, then amplify its beauty beyond your comprehension and imagine what it would be like without death, disease, or any curse upon it, you may have an inkling of heaven.

The word translated "new" in Revelation 21:1 means "new in kind or order" as distinguished from merely new in point of time. Peter describes the process the Lord will use in renovating the heaven and earth that now exist: "But the day of the Lord will come like a thief, in which the heavens will pass away with a roar and the elements will be destroyed with intense heat, and the earth and its works will be burned up. Since all these things are to be destroyed in this way, what sort of people ought you to be in holy conduct and godliness, looking for and hastening the coming of the day of God, on account of which the heavens will be destroyed by burning, and the elements will melt with intense heat. But according to His promise we are looking for new heavens and a new earth, in which righteousness [only] dwells" (II Peter 3:10–13 NASB).

The word translated "elements" is *stoicheiov,* which

means the most basic element of nature. Today we know the atom is the smallest building block of nature. Now Peter says that these elements will be "destroyed." The literal meaning of the word "destroyed" is "to loose something." It was frequently used for untying a rope, or a bandage as in John 11:44. In other words, Christ is going "to loose" the atoms of the galaxy in which we live. No wonder there will be a great roar and intense heat and fire. Then Christ will put the atoms back together to form a new heaven and earth, in which only glorified persons without their sinful natures will live. There will be no more rebellion of man's will against God; only righteousness, peace, security, harmony, and joy.

That's where we want to be!

Go your way, Daniel, for the meaning of the words of prophecy is shut up and sealed until the time of the end. Many shall travel about and knowledge shall increase. None of the wicked shall understand these words, but those who are wise shall in the last days understand.

GOD TO DANIEL
ABOUT 520 B.C.
(LITERAL TRANSLATION)

14

POLISHING THE CRYSTAL BALL

No other prophet has ever had more numerous, far-reaching prophecies revealed to him than Daniel; and yet a great deal of what he received in God-given visions and revelations he didn't understand. Many of his predictions had to do with "the end times" or the events immediately preceding the coming of the Messiah-Jesus to set up God's kingdom.

At the close of Daniel's written account of his prophecies, he reveals his bewilderment about when and how all the things he had predicted would occur: "How long shall it be to the end of these wonders?" (Daniel 12:6 Amplified).

A special angelic messenger replied to Daniel and said, ". . . when the shattering of the power of the holy people [Israel] comes to an end all these things would be accomplished" (Daniel 12:7).

Daniel replied, "I heard, but I did not understand." Then he said, "O my lord, what shall be the issue and final end of these things?" (Daniel 12:8 Amplified).

The messenger replied, "Go your way, Daniel; for the words [of prophecy] are shut up and sealed till the time of the end" (Daniel 12:9 Amplified).

In other words, God revealed to Daniel that his prophecies would not be clearly understood *until* the end times, i.e., the times when the events predicted would begin to take shape. The key that would unlock the prophetic book would be the current events that would begin to fit into the predicted pattern.

Christians after the early second century spent little time really defining prophetic truth until the middle of the nineteenth century. Then there seems to have been a great revival of interest in the prophetic themes of the Bible. Today, Christians who have diligently studied prophecy, trusting the Spirit of God for illumination, have a greater insight into its meaning than ever before. The prophetic word definitely has been "unsealed" in our generation as God predicted it would be.

Now I am about to walk into the lion's den. Perhaps it would be wise to follow Churchill's tactic when he said, "You know I always avoid prophesying beforehand, it is much better policy to prophesy after the event has already taken place." However, in this chapter I will make a number of forecasts about the future which are based on a careful study of the prophetic truth and the writings of many scholars on the subject. I believe that these forecasts are based upon sound deductions; however, please don't get the idea that I think that I am infallibly right in the same way that a Biblical prophet speaking under the direct inspiration of God's Spirit was. I believe that God today gives us illumination to what has been written, but that He doesn't give us infallible revelation as He gave the authors of the Bible. Here, then, are the things that I believe will happen and develop in the near future.

The Religious Scene

In the institutional church, composed of professing Christians who are in many cases not Christian, look for many things to happen:

With increasing frequency the leadership of the denominations will be captured by those who completely reject the historic truths of the Bible and deny doctrines which according to Christ Himself are crucial to believe in order to be a Christian. In some of the largest Protestant denominations this has already taken place. The few remaining institutions which are not yet dominated by the disbelievers will go downhill in the same manner.

There will be unprecedented mergers of denominations into "religious conglomerates." This will occur for two reasons: first, most denominations were formed because of deep convictions about certain spiritual truths. As more of these truths are discarded as irrelevant because of unbelief in Biblical authority, there will be no reason to be divided. Unity is certainly important to have, but never, according to the teachings of Christ, at the expense of the crucial truths of Christianity.

Secondly, as ministers depart from the truths of the Bible they lose the authority and power that it has to meet real human needs, and as many ministers are not truly born spiritually themselves and are consequently without the illumination of God's Spirit, they no longer will be able to hold their present congregations, much less attract others. So they resort to "social action gimmicks," super-organization, and elaborate programs as a substitute.

As Paul predicted concerning these ministers in the last days: "holding to a form of godliness [literally religion], although they have denied its power. . . ." And again he says, ". . . they are always learning and never able to come to the knowledge of the truth" (II Timothy 3:5, 7 NASB).

Young people will continue to accelerate their exodus from the institutional churches. Several surveys taken by church leaders indicate this. Youth today reject impersonal, highly structured organizations with their emphasis upon buildings and material affluence. In talking with many young people from various backgrounds I have found that the institutional churches are viewed by them as a reflection of all they despise in what they consider materialistic, hypocritical, and prejudiced elements within our American culture.

Above all, young people want a simple, personal, and relevant answer to life that isn't based upon self-centered materialism, but upon real life, selfless love. When they are shown that this idealistic view of life cannot be achieved by various shades of welfarism, socialism, or drugs, but only through a personal relationship with Christ that is not tied to joining an institutional church (or religious country club as they call it), then many respond and receive Jesus Christ.

Some traditional churches have learned to provide the personal ministry of God's truth to the youth and you will find their youth departments are flourishing. But they are the exception, unfortunately. Most churches seem to be on the wrong wave length altogether. Some have the truth, but can't communicate it to today's youth; others simply don't teach the truth, and though they try "underground church" approaches, they can't compete with the radical political organizations.

Many youth are going to be on the front edge of a movement toward first century-type Christianity, with an emphasis upon people and their needs rather than buildings and unwieldy programs.

There will be an ever-widening gap between the true believers in Christ and those who masquerade as "ministers of righteousness." I believe that open persecution will soon break out upon the "real Christians," and it will come from the powerful hierarchy of unbelieving leaders within the denominations. Christians who believe in the final authority of the Bible, salvation through the substitutionary atonement of Christ alone, the deity of Jesus Christ, etc., will be branded as prime hindrances to "the brotherhood" of all men and the "universal Fatherhood of God" teaching, which is so basic to the "ecumaniacs" who don't believe in the very heart of Jesus' teachings. Jesus taught that God is the Creator of all, but the Father of only those who believe in Him (John 8:44; Galatians 3:26).

Because of the persecution of believers, there will grow a true underground church of a believing remnant of people.

Look for vast and far-reaching movements toward a one-world religious organization, spearheaded mostly by the unbelieving leaders of the institutional churches; also look for this movement to become more politically oriented than it is now.

Look for movements within Israel to make Jerusalem the religious center of the world and to rebuild their ancient Temple on its old site.

The Political Scene

Keep your eyes on the Middle East. If this is the time that we believe it is, this area will become a constant source of tension for all the world. The fear of another World War will be almost completely centered in the troubles of this area. It will become so severe that only Christ or the Antichrist can solve it. Of course the world will choose the Antichrist.

Israel will become fantastically wealthy and influential in the future. Keep your eyes upon the development of riches in the Dead Sea.

The United States will not hold its present position of leadership in the western world; financially, the future leader will be Western Europe. Internal political chaos caused by student rebellions and Communist subversion will begin to erode the economy of our nation. Lack of moral principle by citizens and leaders will so weaken law and order that a state of anarchy will finally result. The military capability of the United States, though it is at present the most powerful in the world, has already been neutralized because no one has the courage to use it decisively. When the economy collapses so will the military.

The only chance of slowing up this decline in America is a widespread spiritual awakening.

As the United States loses power, Western Europe will be forced to unite and become the standard-bearer of the western world. Look for the emergence of a "United States of Europe" composed of ten inner member nations. The Common Market is laying the groundwork for this political confederacy which will become the mightiest coalition on earth. It will stop the Communist take-over of the world and will for a short while control both Russia and Red China through the personal genius of the Antichrist who will become ruler of the European confederacy.

Look for the papacy to become even more involved

in world politics, especially in proposals for bringing world peace and world-wide economic prosperity.

Look for a growing desire around the world for a man who can govern the entire world.

Look for some limited use of modern nuclear weapons somewhere in the world that will so terrify people of the horrors of war that when the Antichrist comes they will immediately respond to his ingenious proposal for bringing world peace and security from war. This limited use could occur between Russia and China, or upon the continental United States.

On the Sociological Scene

Look for the present sociological problems such as crime, riots, lack of employment, poverty, illiteracy, mental illness, illegitimacy, etc., to increase as the population explosion begins to multiply geometrically in the late '70's.

Look for the beginning of the widest spread famines in the history of the world.

Look for drug addiction to further permeate the U.S. and other free-world countries. Drug addicts will run for high political offices and win through support of the young adults.

Look for drugs and forms of religion to be merged together. There will be a great general increase of belief in extrasensory phenomena, which will not be related to the true God, but to Satan.

Astrology, witchcraft, and Oriental religions will become predominant in the western world.

Where Do We Go From Here?

We believe that in spite of all these things God is going to raise up a believing remnant of true Christians and give one last great offer of the free gift of forgiveness and acceptance in Jesus Christ before snatching them out of the world as it plunges toward judgment.

After considering the incredible things in this book, what should our attitude and purpose be?

First, if you are not sure that you have personally

accepted the gift of God's forgiveness which Jesus Christ purchased by bearing the judgment of a holy God that was due your sins, then you should do so right now wherever you are. It may be that you are bothered because you can't understand it all, or you feel that you don't have enough faith. Don't let either of these things stop you. The only thing you need to understand is that God offers you in Jesus Christ a full pardon and new spiritual life. If you truly desire to receive Jesus Christ into your life, then you have enough faith to enter God's family and change your eternal destiny.

Jesus put the whole thing in a very picturesque way when He said: "Behold, I stand at the door [of your heart] and knock; if any one hears My voice and opens the door, I will come in to him, and will dine [have fellowship] with him, and he with Me" (Revelation 3:20 NASB).

Right at this moment, in your own way, thank Jesus for dying for your sins and invite Him to come into your heart. The door of the above illustration is your desire and will. You open the door by inviting Jesus Christ into your life.

Did you do it? If you did, then where is Jesus Christ right now? According to His promise (and He can't lie), He is in your heart.

Jesus further promised, "I will never desert you, nor will I ever forsake you" (Hebrews 13:5 NASB).

He has come into your heart to stay and to bring new purpose, peace of mind, stability, in spite of circumstances, and true fulfillment to your personality as well as eternal life.

Secondly, if you have received Jesus Christ as Savior, then He wants to change your desires about life to God's desires, and empower you to live for God. God doesn't want us to try to clean up our own lives, but rather to be available to His Spirit who now lives personally within us.

As we trust Christ to deal with our temptations and to work in us His will, then He produces a real life righteousness in us which is characterized by an unselfish love for God and for others. The more we learn of God's love and unconditional acceptance of us, the more we want to please Him and the more we are able to trust Him to work in us. We also will desire to learn His Word which renews our minds to His viewpoint.

Ask Christ to teach you God's Word as you study it and He will.

Third, far from being pessimistic and dropping out of life, we should be rejoicing in the knowledge that Christ may return any moment for us. This should spur us on to share the good news of salvation in Christ with as many as possible. The Holy Spirit is working upon men in a dramatic way and He will lead you to people who are ready or who will be shortly if you trust Him.

Fourth, we should make it our aim to trust Christ to work in us a life of true righteousness. We all grow in this, so don't get discouraged or forget that God accepts us as we are. He wants our hearts to be constantly set toward pleasing Him and have faith to trust Him to help us. John said it this way: "See how great a love the Father has bestowed upon us, that we should be called children of God; and such we are. For this reason the world does not know us, because it did not know Him. Beloved, now we are children of God, and it has not appeared as yet what we shall be. We know that, if He should appear, we shall be like Him, because we shall see Him just as He is. And every one who has this hope fixed on Him purifies himself, just as He is pure" (I John 3:1–3 NASB).

Fifth, we should plan our lives as though we will be here our full life expectancy, but live as though Christ may come today. We shouldn't drop out of school or worthwhile community activities, or stop working, or rush marriage, or any such thing unless Christ clearly leads us to do so. However, we should make the most of our time that is not taken up with the essentials.

Right after one of the major passages of the apostle Paul concerning the Rapture, he gave this promise: "Therefore, my beloved brethren, be steadfast, immovable, always abounding in the work of the Lord, knowing that your toil is not in vain in the Lord" (I Corinthians 15:58 NASB).

As we see the world becoming more chaotic, we can

be "steadfast" and "immovable," because we know where it's going and where we are going. We know that Christ will protect us until His purpose is finished and then He will take us to be with Himself. We can "abound in His work" as we trust Him to work in us and know that it is not in vain because He will give us rewards to enjoy forever for every work of faith.

So let us seek to reach our family, our friends, and our acquaintances with the Gospel with all the strength that He gives us. The time is short.

In the early centuries, the Christians had a word for greeting and departing; it was the word, "maranatha," which means "the Lord is coming soon." We can think of no better way with which to say good-by—

MARANATHA!

NOTES

CHAPTER 1

1. *Time,* March 21, 1969.
2. *Family Weekly,* July 14, 1968.
3. *Los Angeles Times,* November 19, 1968.
4. *Time,* January 15, 1965.

CHAPTER 3

1. Douglas MacArthur, *Reminiscences* (McGraw-Hill: New York, 1964).

CHAPTER 4

1. *U.S. News and World Report,* Oct. 30, 1967.
2. William F. Albright, *From the Stone Age to Christianity* (Doubleday & Co.: Garden City, New York, 1946).
3. John Cumming, D.D., *The Destiny of Nations* (Hurst & Blackette: London, 1864).
4. James Grant, *The End of Things* (Darton & Co.: London, 1866).
5. David L. Cooper, *When Gog's Armies Meet the Almighty in the Land of Israel* (Biblical Research Society: Los Angeles, 1940).
6. Arthur W. Kac, M.D., *The Rebirth of the State of Israel* (Marshall, Morgan and Scott: London, 1958).

CHAPTER 5

1. Walter Chamberlain, *The National Resources and Conversion of Israel* (London, 1854).
2. John Cumming, D.D., *The Destiny of Nations* (Hurst & Blackette: London, 1864).
3. Chamberlain, op. cit.
4. Louis Bauman, *Russian Events in the Light of Bible Prophecy* (The Balkiston Co.: Philadelphia, 1952).
5. Cumming, op. cit.
6. Wilhelm Gesenius, D.D., *Hebrew and English Lexicon.*
7. C. F. Keil, D.D. and F. Delitzsch, D.D., *Biblical Commentary on the Old Testament* (Eerdmans Publishing Co.: Grand Rapids, Michigan).
8. Gesenius, op. cit.

9. Cumming, op. cit.
10. Gesenius, op. cit.
11. Gesenius, op. cit.
12. W. S. McBirnie, *The Coming Decline and Fall of the Soviet Union* (Center for American Research and Education: Glendale, California).
13. Robert Young, LL.D., *Young's Analytical Concordance* (Eerdmans Publishing Co.: Grand Rapids, Michigan).
14. Gesenius, op. cit.

CHAPTER 6

1. *Current History*, "Nasser's Egypt," Gordon H. Torrey, May, 1965.
2. Ibid.
3. *U.S. News and World Report*, July 22, 1968.
4. *Santa Monica Evening Outlook*, December 3, 1968.
5. *Los Angeles Times*, July 9, 1969.

CHAPTER 7

1. J. A. Seiss, *The Apocalypse* (Zondervan Publishing House: Grand Rapids, Mich., 1962).
2. Cumming, op. cit.
3. Ibid.
4. Victor Petrov, *China: Emerging World Power* (D. Van Nostrand Co., Inc.: Princeton, N.J., 1967).
5. W. Cleon Skousen, *The Naked Communist* (The Ensign Publishing Co.: Salt Lake City, 1961).
6. *Quotations from Chairman Mao Tse-tung*, "Problems of War and Strategy" (November 6, 1938), Selected Works, Vol. 11, p. 224. Quoted from a research pamphlet of the Center for American Research and Education, Glendale, Calif.
7. *Bulletin of Atomic Scientists*, "China's Nuclear Options," Michael Yahuda, February, 1969.
8. .... David Inglis, February, 1965.
9. Petrov, op. cit.

CHAPTER 8

1. Merrill F. Unger, *Introductory Guide to the Old Testament* (Zondervan Publishing House: Grand Rapids, Michigan, 1965).
2. Dr. E. J. Young, *The Prophecy of Daniel* (Eerdmans Publishing Co.: Grand Rapids, Michigan, 1949).
3. Sir Robert Anderson, *Daniel in the Critic's Den* (New York, n.d.).
4. *Look*, November 26, 1968.

5. *New York Times Magazine,* May 19, 1968.
6. *Nation's Business,* November, 1966.
7. "Trade and Atlantic Partnership," Dept. of State Publication 7386, Secretary of State Dean Rusk at the Conference on Trade Policy, Washington, D.C., 1962.
8. *Time,* July 4, 1969.

CHAPTER 9

1. William Barclay, *The Revelation of John* (Westminster Press: Philadelphia, 1960).
2. Ibid.
3. *U.S. News and World Report,* August 25, 1969 (source: F.B.I.).
4. *Los Angeles Times,* May 25, 1969.
5. *Columbus Dispatch,* August 21, 1969.
6. *U.S. News and World Report,* November 6, 1967.
7. *Natural History,* May, 1968.
8. *Los Angeles Times,* May 28, 1969.

CHAPTER 10

1. Henry H. Halley, *Halley's Bible Handbook* (Zondervan Publishing House: Grand Rapids, Michigan).
2. *Chimes,* October, 1968.
3. *Los Angeles Times,* May 18, 1969.
4. *Los Angeles Times,* May 25, 1969.
5. *U.S. News and World Report,* July 25, 1966.

CHAPTER 12

1. Bertrand Russell, *Has Man a Future?* (Simon & Schuster: New York, 1962).
2. *U.S. News and World Report,* December 25, 1967.
3. Douglas MacArthur, *Reminiscences* (McGraw-Hill: New York, 1964).
4. Harry Rimmer, *The Coming War and the Rise of Russia* (Eerdmans Publishing Co.: Grand Rapids, Michigan, 1940).
5. J. A. Seiss, *The Apocalypse* (Zondervan Publishing House: Grand Rapids, Michigan, 1962).
6. *Los Angeles Times,* August 20, 1969.
7. *Human Events,* August 26, 1967.

ABOUT THE AUTHORS

HAL LINDSEY was born in Houston, Texas, and educated at the University of Houston. He lived in New Orleans for five years and worked as the captain of a Mississippi River Boat. A graduate of the School of Theology of Dallas Theological Seminary in Dallas, Texas, Mr. Lindsey majored in the New Testament and early Greek Literature. For eight years a staff member of the Campus Crusade for Christ, Mr. Lindsey has travelled to many major universities in the United States and has spoken to tens of thousands of students. Hal Lindsey presently teaches at a Christian Training Center near UCLA with Christian Associates, and lives with his family in Pacific Palisades, California.

CAROLE C. CARLSON is a columnist and free-lance writer whose articles have appeared in a number of newspapers and magazines.